Tool Intelligence as an Explanation of Cross-Linguistic Variation and Family Resemblance

Tool Intelligence as an Explanation of Cross-Linguistic Variation and Family Resemblance

An Evolutionary and Typological Investigation

Anneliese Kuhle

LEXINGTON BOOKS
Lanham • Boulder • New York • London

Published by Lexington Books
An imprint of The Rowman & Littlefield Publishing Group, Inc.
4501 Forbes Boulevard, Suite 200, Lanham, Maryland 20706
www.rowman.com

6 Tinworth Street, London SE11 5AL, United Kingdom

Cover Illustration: Stylized photograph of the *Geschwisterbuche* 'Sibling Beech' in the woodland near Göttingen, Germany (phonetic transcription of the German pronunciation: / g ə ʃ v ɪ s t ɑ b u x ə /). The German name for the two beech trees, which have grown into one another, is a lexical composition of the words for 'sibling' (*Geschwister*) and 'beech' (*Buche*). The modifying element ('sibling') is thus used metaphorically to denote an instance of spatial reciprocity ('two trees connected to one another like siblings'). Photo by Anneliese Kuhle, 2017.

Epigraph: Wolfgang Köhler, *The Place of Value in a World of Facts*, p. 30. Copyright 1966, 1938 by Liveright Publishing Corporation.

Merlan and Rumsey: Merlan, F., Rumsey, A. (1989). *Ku Waru – English word list.* Unpublished manuscript.

British Library Cataloguing in Publication Information Available

Library of Congress Cataloging-in-Publication Data Is Available

ISBN 978-1-4985-6121-1 (cloth: alk. paper)
ISBN 978-1-4985-6122-8 (electronic)

∞™ The paper used in this publication meets the minimum requirements of American National Standard for Information Sciences—Permanence of Paper for Printed Library Materials, ANSI/NISO Z39.48-1992.

Printed in the United States of America

To Sabine Kuhle,
my mother and mentor

Misleading and therefore untrue "correct" information will result when we make statements about unessential phases of an object, while such further knowledge as refers to the fundamental properties of the same object is never added.

—Wolfgang Köhler, *The Place of Value in a World of Facts*

Contents

List of Figures and Tables

FIGURES

TABLES

Abbreviations and Conventions

The following abbreviations are used in the glosses of language material presented throughout chapters 5, 6, 7, 8, and the appendices:

1	first person
2	second person
3	third person
A	transitive subject
Abs	absolutive
Acc	accusative
Adj	adjunct
Aug	augmentative
Ben	benefactive
Caus	causative
Coll	collective
Com	comitative
Dat	dative
Def	definite
Du	dual
E: Nasal	epenthetic nasal
Erg	ergative
Excl	exclusive
F	feminine
Fact	factitive
Foc	focus
Fut	future
Gen	genitive
Group	group suffix
Imp	imperative

Inch	inchoative
Intr	intransitive
Ipfv	imperfective
M	masculine
Min	minimal number
Neg	negation
NF	non-final
Nom	nominative
Perf	perfect
Pl	plural
Poss	possessive
Pres	present tense
Prog	progressive
Pst	past tense
Purp	purposive
Rdp	reduplication
Recip	Reciprocal
Refl	reflexive
Rel	relative
RPst	remote past
RR	reflexive-reciprocal
S	intransitive subject
Sg	singular
Tr	transitive

In addition to the abbreviations listed above, the following conventions are used:

-	separates separate morphemes within one word
=	a clitic boundary between a bound clitic and a free form
.	separates the multiple English words that are used to gloss monomorphemic words in a language
*	ungrammatical form, or: historically reconstructed form
<	indicates that a form derives from another form
/ /	enclose phonological units
[ ]	enclose constituents
‘ ’	indicate English translations of transcribed language material
( )	indicate optional or additional material
/	marks clause boundaries in spoken discourse
^	concatenation: indicates that the connected items occur in strict order
***	indicates an endnote in Table 11.4, Appendix B

For details on notational conventions of language examples, see chapter 4, endnote 1.

Preface

The modern language sciences continue to struggle with conclusive answers to questions that deal with natural languages from a cognitive and evolutionary perspective, not to mention the relationship between language and thought. One possible reason for this struggle is that large-scale interdisciplinary efforts to establish methodological links between specific subfields of linguistic inquiry (e.g., cognitive linguistics, psycholinguistics, sociolinguistics, language typology) and nonlinguistic sciences (e.g., cognitive science, evolutionary biology, comparative psychology, social and cultural anthropology) remain scarce. Many linguists do not know enough about research findings in other fields to support their claims that significant correlations exist between their own findings and the ones offered within the social, behavioral, and cognitive sciences. In other cases, language and nonlinguistic human behaviors cannot be correlated with one another in the ways conceived, although some highly original and successful attempts to establish links between language and thought have been made (e.g., see Lucy, 1992, 1996a; Levinson, 2003; Levinson & Wilkins, 2006; Boroditsky, 2001; Boroditsky & Gaby, 2010). Overall, however, the question of how the connection between language and other domains of cognition and behavior works still remains controversial and awaits further research.

One major obstacle in the current debate on language evolution is the considerable gulf between different theories of language as represented by generative and functionalist theories. Indeed, the discrepancy between the two camps could not be greater and in many cases a dialogue and an exchange of ideas does not occur. To generative linguists, natural languages are anchored in a *modular* cognitive system whose structural (grammatical) features are explicable on the basis of an essentially symbolic mode of information transfer. In this view, it is by no means obvious why one should assume any close links between the ontology of grammatical devices and nonlinguistic capabilities or patterns of human behavior, since in this view, grammar is not determined by either semantic or pragmatic criteria (i.e., meaning). To functionalist linguists, on the other hand, natural languages are not modular but *distributed* in terms of their cognitive foundations in the human individual. The individual's capabilities are here described as being only of a general nature and grammar, in turn, is considered an emergent phenomenon resulting from the functionally oriented interaction between people in coordinated activity and their engaging in intentional communication. In the traditional lingo, generativists emphasize individual biological *competence*, while functionalists put emphasis on actual *performance* in ongoing dialogic interaction. As a result, functionalists detect a myriad of possible ways in which social and cultural values and behavioral conventions may influence and shape the emergence of linguistic patterns of usage, including processes of grammaticalization.

The present work tackles the larger language debate from two major directions. The first is more methodological, the second more theoretical in nature (if it is at all possible to separate the two, theory and method). With respect to the first aspect, methodology, this work assumes a language-typological perspective by assembling and comparing data on a number of individual languages in order to carve out meaningful patterns of variation. This is achieved with a particular focus on reciprocal constructions, that is, a specific functional category of expressions that is found to be pervasive across languages around the world (Maslova & Nedjalkov, 2013). Reciprocal constructions are generally defined by linguists as formally dedicated means of expression that relate to more or less symmetrical or mutual event configurations involving a plurality of participants. The prototypical reciprocal situation involves two or more participants, for example, Mary and Peter, and a prototypical reciprocal construction like *Mary and Peter love each other* (with the reciprocal marker *each other*) contrasts with unidirectional, nonreciprocal constructions like *Mary loves Peter* or *Peter loves Mary* (where no symmetrical relationship is implied in either of these sentences alone). Such dedicated reciprocal means of expression were the focus of much cross-linguistic, typological research throughout the late 1990s and early 2000s. A conspicuous finding of such research was that despite their relative infrequency in natural discourse, many languages nevertheless possess such dedicated reciprocals. Especially to functionalists, this raised the question whether their occurrence in languages could be more than a mere expression of "symmetry" and instead be indicative of some universal yet relativistic tendency to encode sociocultural norms of reciprocal behavior (mutual cooperation, exchange relations) in the linguistic conventions of intentional (human-style) communication.

Two international collaborative research projects in particular stand out in the articulation of these working hypotheses and provide an important backdrop for the current investigation. One of these projects, titled "A Typology of Reciprocal Markers," was carried out from 2005 to 2009 in collaboration between researchers from the Utrecht Institute of Linguistics OTS (led by Martin Everaert) and the Freie Universität Berlin (headed by Ekkehard König). Its major linguistic focus was the investigation of symmetric predicates and their formal encoding from a cross-linguistic perspective. Its larger interdisciplinary focus included the possible overlap between the cross-linguistically defined notion of reciprocity and nonlinguistic notions of social reciprocity as investigated in the social sciences and evolutionary biology (König, 2005; König & Gast, 2008a). Building on earlier work by Nicholas Evans and his team (see below), this typological project explicitly invited cross-disciplinary endeavors to motivate processes of grammaticalization in language through nonlinguistic patterns and conventions of human behavior (i.e., not the other way round, asking how language could shape thought; see references mentioned earlier). A thematically related research project, titled "Reciprocals Across Languages," was carried out from 2003 to 2005 in collaboration between the University of Melbourne (headed by Nicholas Evans) and the Max Planck Institute in Nijmegen (research group by Stephen Levinson). Though with a different areal focus on minority languages spoken in the Pacific region (including Aboriginal Australia), the project was based on a similar functionalist motivation to explain patterns of linguistic marking in terms of "social intelligence" or "social cognition" relatable to forms of social reciprocity (Evans et al., 2011a; Majid et al., 2011a). Unlike the Utrecht/Berlin project, the Melbourne/Nijmegen collaboration added a new methodological dimension to typological research on reciprocity by also including a large sample of video-clip material depicting symmetrical, non-symmetrical and a-symmetrical event configurations as standardized visual stimuli to be tested on informants from different languages. In this way, it was possible to test, in particular, the maximal semantic space attributable to reciprocal constructions in individual languages and to create a cross-linguistically comparable "semantic typology" (Evans et al., 2011a).

The present study was born out of these earlier endeavors to widen the horizon of typological investigations of core-grammatical features and to include increasingly cross-disciplinary working hypotheses in the explanation of grammaticalization phenomena. Chapter 4, in particular, acknowledges this intellectual background to my own investigation of reciprocals in cross-linguistic perspective.

As the title of this book makes clear, however, the present research agenda still lies outside the framework of current mainstream typological investigations of reciprocity and other linguistic phenomena in general. This new agenda aspires no less than a complete synthesis of linguistic typological research with *theories of learning* that currently dominate the debate on language evolution and the psychology of language. Surprisingly, this is only rarely achieved in functionalist explanations of linguistic phenomena. Traditionally, generativism alone has occupied the biolinguistic perspective on language, while functionalism was confined to its association with evolutionary psychology (e.g., research by Michael Tomasello) or social and cultural anthropology in the Boasian and Sapir-Whorfian tradition. This split orientation has not yet changed completely, but my work ventures to sketch a cross-disciplinary research agenda that broadens and complements the functionalist paradigm in a significant way.

A comprehensive review of the cognitive assumptions of different learning theories is taken as the starting point of the relevant discussion (chapter 2). I add to this a new perspective by formulating a tool hypothesis (chapter 3) that leads to the formulation of several testable predictions with regard to the cross-linguistic investigation of natural languages. A detailed report on research carried out on learning, tool use, and cooperative interaction among great apes (from Köhler's early studies at the beginning of the last century to recent work by Savage-Rumbaugh, Boesch, Tomasello, Whiten, and others) offers a new perspective on language learning and the linguistic competence of humans. Basic processes, as for instance recontextualization, the context-dependence of tool use and the "recursive" use of means to produce relevant tools, can be found both in learning processes of apes and of humans and it is only in their capacity for empathy and true collaborative interaction that humans seem to differ from the other primates.

The resultant predictions (derived at the end of chapter 3) are subsequently tested on the encoding of reciprocal relations in a number of Australian and Papuan languages. I have not myself carried out fieldwork on reciprocal constructions in any of the languages analyzed in detail, but I have systematically evaluated the available grammars and grammatical sketches compiled for these languages and also studied available dictionaries and collections of texts. The picture presented here hopefully enriches our knowledge of these structures from both the language-specific and cross-linguistic perspective. Fine-grained surveys of individual languages delineate the specific nature of each language. The chapter on family resemblances (chapter 7), in turn, reunites these cross-linguistic differences and identifies some cognitive-behavioral principles universally involved in language use, which help explain why languages are not only diverse but also similar to one another.

I hope to demonstrate that the results of the descriptive explorations given here are perfectly compatible with functionalist explanations of language structure and moreover with the concept of language as a set of communicative tools. Analogies with the tool use of the great apes throw new light on the traditional metaphor of language as an instrument or tool (Gk. *organon*). In accordance with this view of language as an instrument analogous to prelinguistic tool use, I also provide negative answers to two leading questions: there is neither a semantic core of reciprocal constructions revealed by cross-linguistic studies, nor is there a straightforward parallelism between biological concepts of reciprocity and the use and meaning of reciprocal markers in languages (chapter 8).

As was mentioned at the beginning, research in modern linguistics still struggles with answers to questions concerned with the cognitive and evolutionary roots of human language as well as the relationship between language and thought. The present study offers a new perspective on how to approach these questions in a way that first and foremost connects biological theories of learning with theories of language that are tested linguistically, that is, with a focus on natural language data and preferably not with a focus on European languages alone. The *demetaphorization* of the tool concept in favor of a direct continuity between prelinguistic tool intelligence in great apes and the linguistic intelligence in Homo sapiens enables us to conceive of natural languages as historically evolving tool cultures for intentional communication. As such, natural languages represent emergent behavioral phenomena that do not easily submit to the same constraints of nonverbal behaviors. Just like material tools, linguistic devices are not directly shaped by either biological or cultural dispositions, including *social* dispositions of behavior. Their functional properties are more closely linked to the process of communication itself than to the actual physical representation of biological or cultural "themes" that are prominent in individual human societies. This redefinition of natural languages as genuine *tools* of communication, and *communication* alone, may ultimately also redefine the way that linguists, and functionalists in particular, try to conceive of the relationship between language and thought. I do not think that the present work offers a clearly articulated answer to how exactly this relationship must be conceptualized in *positive* terms. However, it goes a long way towards demonstrating that the tool notion provides a conclusive answer with respect to the cognitive foundation of human languages, which, in turn, provides a necessary platform for further explorations of the relationship between language and thought.

Anneliese Kuhle
Göttingen, July 2018

Acknowledgments

I would like to thank Jana Hodges-Kluck for her wonderful encouragement to undertake this book project and her steady support and professional advice throughout the process of publication. I also thank Trevor F. Crowell, Brianna Westervelt, and Jayanthi Chander for their support during the production process.

The subject matter of this book goes a long way back. I owe special thanks to Ekkehard König for his ability to make reciprocals an interesting topic of study for me and his encouragement to pursue interdisciplinary research beyond the field of linguistics. Nick Evans, in turn, provided me with many helpful hints with regard to the study of individual languages spoken in Aboriginal Australia and the Pacific region. His own field research as well as his expertise on reciprocals were both inspiration and helpful guidance to me.

Several people were closely connected to individual parts of this book project. First, I am very grateful to an anonymous reviewer for their constructive review and criticism. It was much appreciated and I hope I could do justice to their requests. Robert Welker did an excellent job converting my original artwork (all figures) into black and white. Furthermore, I would like to thank Alan Rumsey and Francesca Merlan for their generous support with the linguistic analysis of data on Ku Waru, a Papuan language they have studied intensely for many years. Alan's comments on earlier drafts of chapter 5, in particular, were immensely helpful. Randolf Menzel provided me with critical comments on the discussion in chapter 2, and Ekkehard König offered helpful comments with regard to the preface. I feel that with each revision I was able to improve the clarity of my argument. All remaining errors are my own.

More generally, I owe special thanks to Ferdinand von Mengden for his continued support, professional and friendly advice, and many interesting discussions throughout my postdoc.

This book is dedicated to my mother Sabine Kuhle. Not only do I owe to her the pleasure of being alive, but also the bittersweet awakening that real achievements come from an intricate mixture of effort and ease. I don't think I am fully accomplished in this art yet, but writing this book certainly involved more joy than strain, which I take as a good sign. May the pleasure of reading its lines be a measure of how far I have come in following my mother's advice.

Chapter 1

Languages and Tools

Do language use and the use of tools share a common cognitive basis and does this help us *explain* how languages are of cultural origin? This question is what this book is about and what it seeks to answer in the affirmative. It makes a case for the tool hypothesis. In the passages below, we briefly delineate the argument from other positions in the field. In subsequent chapters, we enter into a theoretical discussion of different conceptions of learning and conduct an empirical, language-typological investigation of *reciprocal constructions* to test our hypothesis. We conclude that languages are genuine "tools of communication" and thereby argue in favor of cognitive continuity rather than discontinuity with respect to prelinguistic behaviors in nonhuman animals. In particular, we draw attention to conspicuous similarities between natural languages in human societies and the beginnings of material cultures in nonhuman primates.

LANGUAGES, THOUGHT, AND CULTURE

Current linguistic theorizing shows a strong divide between biolinguistic and usage-based explanations of language evolution, historical development, and its cognitive underpinnings (Ibbotson & Tomasello, 2016; Berwick & Chomsky, 2016). The biolinguistic, generative perspective embraces the idea of genetic endowment for the acquisition of linguistic structures, such as the use of passives like English *I was bitten by a dog*, predications in past tense like German *Sie ging gestern zum Bäcker* 'she went to the bakery yesterday' or the expression of reciprocal meanings like French *Ils s'embracent l'un l'autre* 'they embrace one another.' Such universal, human-specific predispositions for linguistic behavior predict that cross-linguistic variation is only superficial and that universal structural categories persist throughout the languages around the world. Take, for instance, the following quotation from Berwick and Chomsky (2016) in which the Principles and Parameters (P&P) approach, one of the most influential versions of the biolinguistic theory, is offered as an explanation of cross-linguistic variation:

> The P&P approach is based on the assumption that languages consist of fixed and invariant principles connected to a kind of switchbox of parameters, questions that the child has to answer on the basis of presented data in order to fix a language from the limited variety of languages available in principle [. . .]. For example, the child has to determine whether the language to which it is exposed is "head initial," like English, a language in which substantive elements precede their objects, as in *read books* or whether it is "head final," like Japanese, where the counterparts would be *hon-o yomimasu*, "books read." As in the somewhat analogous case of rearrangement of regulatory mechanisms, the approach suggests a framework for understanding how essential unity might yield the appearance of the limitless diversity that was assumed not long ago for language (as for biological organisms generally). (Berwick & Chomsky, 2016, pp. 68–69)

As we will see in chapter 2, this *faculty of language approach* is based on a very specific tradition of theorizing in evolutionary developmental biology (*evo-devo* research). It rests on a switch box metaphor associated with genetic regulatory mechanisms and embryogenesis of organisms more generally. Even so-called adaptationists among generative linguists do not concur with this line of reasoning and prefer to see language grammars in true analogy to complex biological systems like vision, echolocation or spatial navigation (Pinker & Bloom, 1990; Pinker & Jackendoff, 2005; Hurford, 2012). However, it remains to be seen whether the degree of variation in behavior fixed by biological systems equals that of cross-linguistic diversity. Moreover, this idea of an evolved language instinct has the disadvantage that it cannot properly explain the *symbolic* character of language use, that is, the fact that linguistic patterns are not physically directly correlated with the things they refer to in verbal discourse.

According to proponents of the usage-based, functionalist or cultural view of language, the linguistic evidence for absolute cross-linguistic structural universals is scarce (Evans & Levinson, 2009; Everett, 2012; Ibbotson & Tomasello, 2016). The more languages one studies, the more linguists 'detect new categories that do not fit existing taxonomies' (Haspelmath, 2007a). Thus, arguably, the comparisons between languages simply do not fit the idea of pre-existing, innate constraints with regard to the grammatical structures of natural languages. Along these lines, Tomasello (2008) remarks:

> What we need is an account that will enable us, in the end, to explain the emergence *not* of Language (with a capital L), but the emergence of 6,000

> different human languages with 6,000 different sets of communicative conventions—including grammatical conventions for structuring multiunit utterances into coherent messages. Of course there will be universals—it is everywhere the same species with the same cognitive tools trying to do many of the same things—but there are particularities as well, and indeed the fact that humans do not have one single means of communication for all members of the species is already an evolutionary novelty that must be explained. (Tomasello, 2008, pp. 243–44; emphasis not added)

As we will see in the second part of chapter 2, this *general intelligence hypothesis* rests on modern assumptions about neural plasticity in the late ontogenetic development of organisms. It assumes that general learning abilities exist in many animal species, including invertebrates, and that flexible behavior is based on some amount of context-dependent reasoning rather than either instinct or trial-and-error learning.[1] As is shown in ape language studies, such general reasoning skills are also involved in the acquisition of communicative symbols and the comprehension of spoken language. However, as Sue Savage-Rumbaugh and others have argued, the social surroundings of the subjects are crucial in this process of acquisition. True linguistic skills are only acquired in a socially scaffolded environment where individuals share common grounds of experience and specifically attend to one another while engaging in shared activities.

Over the past decades, this last insight has led to a new position within the general intelligence approach, the so-called social intelligence hypothesis. According to this view, humans have evolved special cognitive skills for altruistic and collaborative interaction (Tomasello, 1999, 2008). This in particular has enabled them to take intentional communication past the stage of simple requests and to invent the declarative communicative means of informing, sharing, narrative, and so on, which are characteristic of today's languages. This position has serious consequences for the current debate on language evolution.

THE CURRENT STATE OF THE ART

If the 6,000 languages around the world are indeed the product of cultural-historical development, as the functionalist view maintains, the quest for a universal cognitive basis does not become obsolete. "The issue [. . .] is not whether human beings are biologically prepared for language acquisition; they are" (Tomasello, 1995, p. 137). The issue is whether human beings come equipped with an innate linguistic module (Universal Grammar) or whether their linguistic skills rest on more general and neuroplastic learning abilities that begin to unfold in early childhood. Moreover, the issue is whether the general intelligence hypothesis can make plausible the idea that language evolution started out on the basis of prelinguistic abilities for which we might still find evidence in our closest living relatives, the great apes.

Current research, however, is not drawing anywhere near such a conclusion. This might seem a surprising statement. But really, given the views in the mainstream literature, it is not. Even scholars in general agreement with the claim that general learning abilities underlie the human capacity for speech ("speech" in the sense of language) discard comparative psychologists' positive claims in favor of cognitive continuity rather than discontinuity. The general reception of this position is as follows:

> Tomasello and his collaborators have produced an extensive body of results from experiments with chimpanzees and human children, demonstrating that very young—even preverbal children display an eagerness for social engagement not found in our nearest living relatives. [. . .] Building on this and a long history of findings in social philosophy and developmental psychology, Tomasello et al. have hypothesized that the evolution of shared intentionality was a crucial antecedent to human language and the socially constructed worlds in which we live our cultural lives. (Hawkes, 2012, p. 687)

In other words, even proponents of the social intelligence hypothesis, like proponents of a generative view of language, settle for *special* cognitive abilities not shared by nonhuman primates. Instead of universal grammar, they rely on notions of *shared intentionality* or *human sociality*, involving altruistic motives and theory of mind, in order to explain the human capacity for language (Tomasello, 1999, 2008; Gärdenfors, 2004; Enfield & Levinson, 2006; Dor et al., 2014). Unfortunately, therefore, even the mainstream functionalist view of language currently rests on assumptions of discontinuity rather than continuity in explaining the evolution of language.

Of course, there is no ultimate reason why such an explanatory model should not render the best possible explanation of human intelligence and the existence of natural languages. However, perhaps the last word has not been spoken in this debate. We can't preclude the possibility that there is still further evidence that speaks more in favor of cognitive continuity with prelinguistic behavior than is currently believed. Even Michael Tomasello, despite his skepticism, acknowledges that "it is of course possible that we just do not understand everything well enough to find the hidden simplicity" (Tomasello, 2008, p. 319).

The existence of a "hidden simplicity" is what this book is about. It draws attention to a model of learning in reference to prelinguistic animals, notably nonhuman primates, which arguably bears true analogy to human linguistic behavior. This model draws attention to structural and functional features in language use that are also characteristic of prelinguistic behaviors in nonhuman primates. Such behavior is here identified under the labels of *tool intelligence* and *material cultures*. The evidence is based on criteria of cultural development that are formulated independently of language. We can apply these criteria to natural languages, even grammatical constructions, and verify them with a focus on cross-linguistic variation and family resemblance. In the following, let us sketch how this analogy comes about and offer a perspective why it is important to accommodate this view in current theorizing about language.

TOOLS, THOUGHT, AND CULTURE

Tools are inert objects or complex units of behavior with which to manipulate the environment *intentionally*. Following the primatologist Christophe Boesch, "[a] tool is defined as an external object detached from its substrate use to attain a goal (Goodall, 1970; Beck, 1980). In a sense, a tool is perceived as an extension of the body that is used to achieve a goal that cannot be directly achieved with the use of only hands or teeth" (Boesch, 2013, p. 24).[2] Furthermore, "[f]lexible tool use requires the ability to manipulate physical causal relations in a generalized manner" (Byrne & Russon, 1998, p. 681). "To be able to achieve such a task, an individual needs to have a certain level of causal understanding about the effect of one's own body on objects as well as on the body's actions on external objects" (Boesch, 2013, p. 24). In other words, real tools are not specific to certain contexts of usage. They are generalizable across many different contexts of usage where similar problems arise. They are thereby characterized by *means–ends dissociation*: the same means are used for different ends and different means may be used to achieve the same ends (Call, 2013). It requires an extended awareness or problem orientation that allows a subject not only to seek a solution in relation to a specific goal region, but a problem more generally. It requires human-style intentionality (Malle et al., 2001).

Such tool-using abilities become particularly prominent in nonhuman primates even under natural conditions of living. Chimpanzees use stones or wooden clubs to crack open nuts and other hard-shelled fruits. They use thin sticks to fish termites off the ground; they insert twigs inside beehives to retrieve honey; or they use sticks to scoop algae from water surfaces. Indeed, their tool repertoires are so rich that primatologists working in the field have begun to catalogue ape tool behaviors in their various areas of life and speak of these tool repertoires as *material cultures* (Whiten et al., 1999, 2001, on chimpanzees; Van Schaik & Knott, 2001, and Van Schaik et al., 2003, on orangutans). *Material culture* is the term used because the tool repertoires between different groups differ conspicuously. These cross-group differences are not sufficiently explained by either genetic or ecological factors alone. For instance, wild ape populations in different regions of West Africa rely on different techniques of nut-cracking. The chimpanzees in Taï Forest rely on stones or wooden clubs to crack open nuts on a hard surface like roots or naturally embedded stones. The chimpanzees from Bossou, on the other hand, rely on stones as both hammer and anvil. They recruit flat-shaped stones to place underneath nuts before they start hitting with another stone from above. Moreover, Taï chimpanzees habitually feed on five different nut species using this skill; Bossou chimpanzees, on the other hand, only crack open one type of nut species. Such differences in behavior are not explained by the availability of potential tool objects within the natural habitat; nor are they explained by the availability and nature of food resources. Rather, the only explanation that can plausibly explain the emergence of such arbitrarily different behavioral variants is one in terms of incidental innovation and subsequent social transmission, that is, cultural-historical development (McGrew, 1998; Boesch, 2012).

Thus, the *patterns of variation* in tool use between different groups are conspicuous. Technical skills vary in form and they vary in application. On top of this, they also show traces of *cumulative cultural development* (Boesch, 2013, pp. 39–40). For instance, stick uses are often highly skilled and require various steps of preparation and modification. Individuals collect twigs long before arriving at a feeding site. They select twigs from some plants and not others. They strip raw twigs of their leaves so as to handle them in a special manner. For fishing insects, they hold the twigs close to insects' nests or the trails between nests. They wait several seconds or even longer before retrieving them, moving the twigs in a manner so as to attract more insects. In some communities, subjects sweep off the insects with their mouths, in others they use their hands instead. In other words, even such primitive tool-using behaviors involve many small steps of functional behavior that reflect on their proficiency and foresight. Perhaps more importantly in this context, many of these steps are independent of one another. They must not have been invented in a one-time scenario of problem solving. Different subjects living at different times may have administered different context-dependent changes to these behaviors, which were subsequently picked up by other group members. If this is true, prelinguistic tool uses in nonhuman primates are *constructed, arbitrary behaviors* that are accumulated and constantly renegotiated in many small steps over many generations of tool users.

THE ANALOGY TO LANGUAGE

By analogy, human languages are complex tool repertoires within the domain of intentional communication. Humans rely on verbal means to manipulate the mental states of others. *Manipulation* is here understood in analogy to physical manipulation, that is, the influencing of other entities by means of a certain degree of insight into their states-of-being and potential responses under different external conditions. In the functionalist literature, it is quite common to view language as medium of communication that is used to influence the mental states and attitudes of others (e.g., Tomasello, 2008, p. 343; Gärdenfors, 2004; Kolodny & Edelman, 2018). One important difference, however, is that the tool analogy offers a definition of manipulation as "a medium of communication that involves both positive and negative social encounters." It is thus defined here in a more neutral sense than it is commonly understood in the context of specifically *social encounters*, where manipulation often implies only negative attitudes such as deception or exploitation. It is not so much concerned with the evaluative aspect of the social encounters in which subjects communicate. Rather, it is more focused on the fact that subjects possess insights and *causal understanding* of the mental states of others at all and seek to directly influence these mental states by relying on communicative means as a way to take action. The goals are naturally subjective and based on personal interests of an individual, whether these imply positive or negative outcomes.

Furthermore, in this view of language, linguistic means, like prelinguistic tools, are constructed, arbitrary means that are naturally somewhat detached from the environment, even the specific environment to which they owe their functionality. The reason is that they are *individually and historically invented means of communication*. In the case of language, the functionality is linked to the social environment, that is, the attitudes and common grounds shared with other individuals within the community. The functionalist literature often acknowledges this property in language. One of the most conspicuous and often-cited characteristics of human language is the *arbitrariness of the linguistic sign*. "Communicative conventions are defined by two separable characteristics (Lewis, 1969). First and most critically, we all do something in the same way because that is the way everyone is doing it [. . .]: it is shared. Second, we could have done it differently if we had wanted to: it is, at least to some degree, arbitrary" (Tomasello, 2008, p. 219). Yet, what is not often argued is that this is a conspicuous characteristic that the linguistic sign shares with prelinguistic tool uses. As we have seen earlier, one of the most conspicuous characteristics of intelligent tool use is the phenomenon of *means–ends dissociation*: the same means are used to reach different ends and different means are used to reach the same ends. The very same is also true of natural languages.

Finally, the tool-use analogy predicts that linguistic constructions are not the product of single-step innovations. They are the result of long-term, *cumulative historical developments* that proceed over many generations of language users. In a way, this is another kind of *arbitrariness* often attributed to the linguistic sign. It refers to the fact that the form of a linguistic pattern might not have been assembled in a single step but might have undergone various developments over successive generations of language users. It thus shows many structural and functional features that are in fact the product of a long succession of previous usages rather than a direct (intelligent) response in usage under extant conditions of intentional communication. This is also known as the *ratchet effect* of social transmissions within the group and captures the fact that all innovative steps are built on top of already existing, culturally inherited variants of behavior.[3] Intriguingly, this is a view very much in line with current, usage-based linguistic theorizing embracing the idea of *grammaticalization*. Put very simply, the term captures the idea that grammatical means such as complex sentence types, word morphologies, pronominal expressions, and so forth, are the opaque result of long-term historical developments in individual languages. These start out with the sporadic recruitment of lexical materials or other constructional means under slightly novel conditions of usage and intentionality and are subsequently reinforced and re-conventionalized through frequency effects in everyday discourse (Heine et al., 1991; Hopper & Traugott, 1993; Lehmann, 1995 [1982]; Croft, 2000; Traugott & Dasher, 2002; Hawkins, 2004; Haspelmath, 2008a, 2008b). Of course, due to their complexity, faithful historical explanations of linguistic patterns (constructions) are far more demanding than in the case of prelinguistic tool uses. It seems almost hopeless to compare the two behaviors in this respect. And yet, as this book seeks to demonstrate, there are ways and means to test the tool hypothesis even on natural languages with such a long-lived history.

We can't anticipate the systematic investigation of this book. But we can point out some of the characteristic patterns of variation in cross-linguistic comparison that make natural languages perfect *tools* of communication. For example, natural languages rely on different structural means to encode similar meanings or functions. A classic example is the encoding of grammatical relations like subject and object either by word order or morphological case marking. Some languages, like English, rely on word order to distinguish between different participant roles like subject/agent and object/patient of an action. Others depend less on word order and rely more on case marking on different noun phrases. For instance, in German, subjects are marked by nominative case (*der Hund* the.Nom dog), while objects are marked by the accusative (*den Hund* the.Acc dog). Still other languages, for instance, Papuan and Australian Aboriginal languages, are even less dependent on word order than languages like German. They even allow the scrambling of elements belonging to the same noun phrase (e.g., see Foley, 1986, p. 167 ff.; Hale, 1992; Dixon, 2002, pp. 78–79). Another well-studied example is the structural variation in possessive marking. Minority languages studied in the Pacific region have far more intricate structural encodings of possession than any European language user would be able to anticipate (Lichtenberk, 2009; Aikhenvald, 2013, p. 20 ff.). For instance, in most European languages, 'my orange' would be about the only way in which possession could be expressed. In the Oceanic language Lolovoli (Ambae), by contrast, there exist different forms of specifying different variants of this relation: for example, Lolovoli *ga-ku moli* 'my orange (for me to eat)' versus *me-ku moli* 'my orange (for me to drink)' (Hyslop, 2001, p. 185; cited in Lichtenberk, 2009, p. 274). In other words, some languages show intricate structural encodings where other languages have a "blind spot." The different possibilities of interpretation are a function of different structural encodings, not simply context. Another dimension of variation is the notorious irregularity and differential productivity of morphological markings on different word-class members in a language (Bauer, 2001). There is no way one could predict the productivity of such morphologies across different languages. The spectrum ranges from entirely nonproductive to moderately productive to highly productive. Finally, languages show variation in the etymological origins of their grammaticized forms. A classic example is the study of future markings in different languages and the finding that such markers derive from different lexical source materials, such as verbs of volition ('want, desire' as in English *He will visit me*) or verbs of motion ('go' as in French *Il va dire* 'he is going to say') (Bybee et al., 1994).

Language typologists, who engage in large-scale cross-linguistic comparisons, already realize and argue that their objects of study display all the features of behavioral patterns that are the result of cultural-historical development (Heine et al., 1991; Hopper & Traugott, 1993; Haspelmath, 2007a, 2008a; Croft, 2000; Traugott & Dasher, 2002; Evans & Levinson, 2009; Everett, 2012). However, what language typologists don't realize as yet, for they don't engage in that sort of comparison, is that these patterns of variation are the very same as those which primatologists identify in natural tool-use repertoires among nonhuman primates. The consequences of this new insight have yet to be drawn and incorporated into linguistic theorizing about language evolution and other fields of inquiry.

OVERVIEW

If our assumptions about a cognitive homology between language use and prelinguistic tool use are correct, we should be able to test our hypotheses on natural languages. This is the task of this book and thus of the chapters to come. In chapter 2, we lay out the existing debate and review current models of learning and cognition in some detail. In chapter 3, we deal with the tool model and derive concrete predictions that we subsequently test on *reciprocal constructions* in natural languages. Reciprocal constructions are complex sentences, involving some special verbal morphology, pronominal forms, or other constructional means, which encode complex relations of mutuality or more-or-less symmetry among a plurality of participants (e.g., 'they help / hit / see / like each other'). The interested reader is referred to chapter 4 and following to learn more about this grammatical device and how it is used in different languages around the world. The focus will be on structural variation (chapter 5), functional variation (chapter 6), and family resemblance (chapter 7). Moreover, chapter 8 turns to the question of a possible relationship between linguistic reciprocity and behavioral concepts of social reciprocity. Throughout the discussion, the tool hypothesis poses a central basis for explaining the nature of reciprocal constructions and their relationship to real-world experience. Chapter 9 concludes our findings. In particular, it concludes that we are able to empirically verify cognitive continuity ("hidden simplicity") between linguistic and prelinguistic behavior where otherwise, following current mainstreams, none is seen to exist.

NOTES

1. Interestingly, but also somewhat confusingly, at its base, this latter approach also shares various tenets formulated by linguists who still embrace the adaptationist version of Universal Grammar. For instance, in his two impressive volumes on *The Origins of Meaning* (2007) and *The Origins of Grammar* (2012), James Hurford sheds light on various aspects of the evolution of language. At first, in order to explain fundamental psychological abilities involved in language use and comprehension, he draws close analogies to prelinguistic animals and their grasp on conceptual knowledge. Yet, in the end, his generative assumptions prevail and let him settle for a view similar to that of Jackendoff and others. A major reason seems to be that linguistic structures, that is, words, phrases, sentences, are here still conceptualized as structures with an abstract, syntactic, symbolic form that exists independently of the conceptual abilities of an individual. According to Hurford, at least, the use of such complex linguistic patterns in discourse does not resemble anything one encounters in the behaviors of prelinguistic animals.

2. Others, including Richard Byrne, add that simple elements of behavior, too, can function as "tools," given that they are recombined with other elements of behavior for novel purposes for which they were not previously used (Byrne & Russon, 1998, pp. 674–75).

3. The term *ratchet effect* is taken from Tomasello (1999) and Tennie et al. (2009). In the present work, it is used in analogy to prelinguistic tool uses. Tomasello and colleagues seek to apply the term to human culture alone.

Chapter 2

Language in the Light of Biological Evolution

Linguists and psychologists alike try to bridge the gap between studies in modern linguistics and studies in modern psychology, behavioral biology, and cognitive science.[1] All such interdisciplinary endeavors have in common that they recognize certain features of natural languages as essential and, moreover, as requiring an explanation in terms of biological evolution by natural selection. Most experts concur that human language is a behavior like other nonverbal human and nonhuman behaviors. That it fulfils a particular function and that it has an evolutionary history, which is specific to the human species. As a result, one also concurs that humans possess some sort of learning disposition that enables them to acquire linguistic behavior. Moreover, one agrees that linguistic activity is characterized by symbolic reference, intentionality, and a high degree of structural complexity. It is based on complex propositional reasoning and is practiced under displaced conditions of action.

Yet, experts do not agree on the explanatory foundation. In a sense, language in the light of biological evolution is still an unsolved mystery. Two very different conceptions of human language are currently most prominent in linguistic theorizing: language as an *instinct* and language as an expression of *general intelligence*. The former conception follows the tradition of the nativist turn pioneered by the American linguist Noam Chomsky and is commonly known as the generative theory of language or the *faculty of language hypothesis*. It proposes an innate faculty of language whose initial state is referred to as Universal Grammar (UG). The other conception runs in the tradition of cultural anthropology (Boas, Sapir), historical linguistics (von Humboldt; von der Gabelentz), and Gestalt psychology (Bühler; Vygotsky). It is commonly known as *usage-based approach* (Ibbotson & Tomasello, 2016), or else the functional (cultural) theory of language. It postulates a general intelligence in humans that is specific to physical and social interaction more generally, not linguistic behavior in particular.

Needless to say, the literature offers many different takes on each theory of language and it is impossible to do justice to all of them. Yet, it is possible to carve out the essential premises of each theory and to reflect some of their consequences.

THE FACULTY OF LANGUAGE HYPOTHESIS

The basic assumption that humans must have an innate cognitive capacity, a faculty of language (FL), to learn and speak a natural language was pioneered by the American linguist Noam Chomsky (Chomsky, 1959, 1965, 1980, 1981). The theory of FL was meant to explain two things: First, it answered the question of why natural languages have many structural features in common (the linguistic universals argument). And second, it explained why young children are capable of acquiring a language within a relatively short period of time, during a *critical period*, despite their limited exposure to primary linguistic data (the poverty of the stimulus argument). Chomsky (2009) recalls how intellectual developments in the mid-twentieth century led to this emergence of a "biolinguistic perspective" on language, whose "real object of inquiry" became "the internal mechanisms that generate linguistic expressions and determine their sound and meaning" (Chomsky, 2009, p. 16).[2] This development of a theory of generative grammar was a response to radically empiricist theories of learning, which back then dominated research in American psychology (Cowie, 2017 [2008], chapter 1, pp. 4–9; Samet & Zaitchik, 2017 [2012], chapter 1.1, pp. 4–11). The significance of this counter-movement to behaviorism, which is famously associated with Chomsky's (1959) criticism of B. F. Skinner's book *Verbal Behavior* (1957), did not so much reside in its criticism of anti-mentalism per se, as in the particular *theory of nativism* (mentalism) that was set against it. This shift away from behaviorism, which later became known as the *nativist turn*, consisted in the adoption of a view of cognition that was specifically informed by models of *symbolic computation* (Samet & Zaitchik, 2017, chapter 3.1, pp. 43–47; Garson, 2012 [1997]).

> According to this view, the brain incorporates "specialized organs," computationally specialized to solve particular kinds of problems, as they do with great facility, apart from "extremely hostile environments." The growth and development of these specialized organs, sometimes called "learning," is the result of internally directed processes and environmental effects that trigger and shape development. The language organ is one such component of the human brain. (Chomsky, 2002a, p. 64)[3]

The generative notion of language as an "instinct," "organ," or "module" thus combines two essential premises: First, that the initial mental state of the language system (= universal grammar) is not a blank slate but a structurally prespecified area of the brain dedicated to the processing of primary linguistic data, which is analogous to the initial states of other cognitive systems. And second, that the structural development of this initial state into a full-fledged language system (= faculty of language) involves processes of growth and maturation ("sometimes called 'learning'"), which result in an attained state that relies on a special form of symbolic (linguistic) representation.

The Non-Adaptationist Approach

Chomsky's theory of language underwent several reformulations throughout the 1960s to 1980s, often initiated by Chomsky himself (Chomsky, 1965, 1975, 1980, 1981, 1995), which led to some important changes in the conceptualization of the postulated FL (Cowie, 2017). From the cognitive perspective, a major concern of these restatements related to the precise nature of the *initial state* of the system. The particular challenge was to explain how, by what process of learning, a system of *symbolic* representation could be biologically implemented and acquired. The special feature characterizing systems of symbolic representation, such as digital computers, is that they are not structurally (physically) determined by the semantics that they represent (Garson, 2012). The implementation of syntactic rules is thus independent of the specific semantic variables on which they operate symbolically (Samet & Zaitchik, 2017, chapter 3.1, pp. 43–47). The special feature of learning by experience, on the other hand, is that the mental representations resulting from it are structurally determined by the sensory inputs (semantics) that are processed by the system. Hence, the biological notion of "representation" and "learning" is invariably built on the assumption that the transition from an initial (naïve) to an attained (competent) state is dependent on structural adaptations of the system in response to external (sensory) stimulation. As a consequence, the representations of natural biological systems are taken to be *adaptive* or functional, not structurally *non-adaptive* or symbolic.

The functional-adaptive acquisition of language, however, was explicitly rejected by the nativist turn. The rejection is commonly known as the *poverty of the stimulus argument*. The essence of this argument is that the acquisition of language cannot be explained on the assumption of general (empiricist-style) principles of learning, because the primary linguistic data (pld) to which children are exposed during early childhood is not prestructured enough to convey the regularities (rules) that govern a language (Cowie, 2017, chapter 2.2, pp. 16–35). Hence, true linguistic competence cannot be acquired through experience by the child and has therefore to be innately prespecified in the form of a priori knowledge. In other words, the basic principles, symbolic rules, from which all other, more specific rules governing individual languages can be generated (derived), are already inherent to the initial native state (universal grammar). However, the structural development of the system—during ontogenetic development—has thereby become completely detached from learning by experience. For how could an innate set of symbolic, structurally invariant, generative rules become *shaped* by experience (Samet & Zaitchik, 2017, chapter 3.1, pp. 43–47)?

In answer to this question, Chomsky proposed the view that the process of development that characterizes language acquisition is in fact not a form of associative learning at all but instead a special form of growth and maturation (Cowie, 2017, chapter 2.1, pp. 9–16; Samet & Zaitchik, 2017, chapter 1.2, pp. 11–16). This learning process is only very indirectly influenced by external stimulation:

> The initial state [universal grammar] changes under the triggering and shaping effect of experience, and internally determined processes of maturation, yielding later states that seem to stabilize at several stages, finally at about puberty. We can think of the initial state of FL as a device that maps experience into state L attained: a "language acquisition device" (LAD). (Chomsky, 2002a, p. 85)

It is easy to mistake the true conception of "learning," which is proposed here. As we have seen earlier, the type of learning clearly rejected by the generative theory is the type of associative learning put forward by empiricists as the basis for structural development in other biological systems. Hence, the alternative to empiricist-style learning rests on a very different conception of structural development. It was chiefly imported from developmental biology and *evo-devo* theorizing. To show this, we consider one of the most influential models of the generative theory developed throughout the 1980s, which is known as the Principles and Parameters (P&P) approach.[4] The solution offered by this approach is to marginalize the influence of experience to "fairly elementary experience" and to introduce the analogy to "regulatory mechanisms" as discussed in developmental biology (see the quotation below).

> This principles and parameters (P&P) approach was based on the idea that the format consists of invariant principles and a "switch-box" of parameters [. . .]. The switches can be set to one or another value on the basis of fairly elementary experience. A choice of parameter settings determines a language. The approach largely emerged from intensive study of a range of languages, but as in the early days of generative grammar, it was also suggested by developments in biology—in this case, François Jacob's ideas about how slight changes in the timing and hierarchy of regulatory mechanisms might yield great superficial

> differences (a butterfly or an elephant, and so on). The model seemed natural for languages as well: slight changes in parameter settings might yield superficial variety, through interaction of invariant principles with parameter choices. (Chomsky, 2009, p. 23)

According to this view, the discovery of so-called deep homologies among organisms proved the existence of a "developmental-genetic toolkit," a kind of universal Bauplan, from which all "superficial variety" of extant organisms derives (Chomsky, 2010, p. 51). The suggested analogy to language is one where the identified deep homologies are the equivalent to Universal Grammar. The different genetic expressions during embryonic development (yielding different organismic body structures), in turn, are analogous to the different expressions of UG, yielding the different grammars of individual languages.

The notion of a "developmental-genetic toolkit" is taken from developmental biology and hence the early ontogenesis of organisms. All humans grow arms and legs and inner organs like liver, heart, and kidney, irrespective of the precise environmental conditions under which they are reared. The same logic applies to the basic architecture of the human brain: all humans grow sense organs like eyes, ears and noses; moreover, their brains show the same organization into different cortical and subcortical areas. Moreover, each of these morphological features can be distinguished to a certain degree from those developed in other species. As is comprehensively reviewed in Deacon (1997), this early structural development (organogenesis) of the human brain is clearly driven by *regulatory mechanisms*, which are based on expressions of the genes and resulting processes of neural growth and maturation (Deacon, 1997, p. 174 ff.). This means that many morphological features of biological systems, including the sensory organs and their neural projections within the brain, are to some extent already in place before an individual starts interacting with its environment, and hence starts to learn by association. The *initial states* of biological systems such as the visual or auditory system thereby rely on processes of *internally determined growth and maturation*. They are an expression of the genes.

By analogy, the notion of a "developmental-genetic toolkit" thus also refers to the structural development that leads to the *initial state* of the language system. Just as the initial (native) state of the human visual system is in place before visual learning takes place, so the initial state of the human faculty of language (= UG) is in place before the actual process of language acquisition sets in. This, however, is not the entire scope of the analogy. It *also* explains the developmental transition from the initial state of the system (UG) to its attained state (FL). Thus, in reference to the actual acquisition process, too, Chomsky and others[5] assume that endogenously determined processes of growth and maturation guide the structural development of the system. In this light, speaking of "environmental effects that trigger and shape development" (Chomsky, 2002a, p. 64) or "the triggering and shaping effect of experience" (Chomsky, 2002a, p. 85) has a very distinct meaning. It does not describe "learning by experience" but developmental processes that bear analogy to early organogenesis (Deacon, 1997).

Of course, what we have ended up giving is the perfect description of a *biological instinct*.[6] What characterizes an "instinct" on the behavioral level is that an individual is able to perform an action in response to external stimulation without or only with little prior experience. It knows how to act without having encountered or witnessed this type of situation before (e.g., gaping response in herring gulls, egg roll behavior in greylag geese, prey-catching behavior in toads, flight coordination in flies, face recognition in human infants, etc.).[7] What characterizes such instinctive behavior on the cognitive level is that the initial (native) state of the system is roughly equivalent to the attained (adult) state of the system. In other words, not much structural development is gained from experience. Hence, the genetically determined processes of growth and maturation can be said to have determined both the initial and the attained state of the system, which is precisely what the generative conception of language acquisition—as we have contemplated it so far—seems to convey.

However, we now face *a language-empirical problem*, which is not compatible with this notion of biological instinct: diversity. This is where eventually the *switch box* or *toolkit* metaphor comes in. The problem is that the grammars of individual languages are not the same. Languages differ with respect to the most basic structural parameter settings (e.g., word order, case marking, analytic versus synthetic, pro-drop). Moreover, from the perspective of symbolic computation, they show many imperfect design features such as zero marking or semantically uninterpretable (redundant) morphology (Chomsky, 2002a, pp. 88–89). We now have to assume that children are able to learn by instinct different grammars based on the limited environmental inputs that they receive. This however is a contradiction in terms. How can an endogenously determined "instinct" in members of the same species explain the development of such diverse attained states?

The solution offered by the P&P approach (as quoted earlier) is to introduce the additional notion of a developmental-genetic switch box or toolkit. This notion, however, does *not* come from insights gained from the study of intraspecies developments of instinctive behaviors. The latter draw on learning by experience to explain the acquisition of different competences in the attained states.[8] Instead, it comes again from studies in developmental biology and is based on evolutionary developmental theorizing (*evo-devo*).[9] It is in the context of such comparative developmental research and the identification of evolutionary genetic conservatism that Chomsky adopted the idea that "slight changes in the timing and hierarchy of regulatory mechanisms might yield great superficial differences (a butterfly or an elephant, and so on)," and then proposed that "[t]he model seemed natural for languages as well: slight changes in parameter settings might yield superficial variety, through interaction of invariant principles with parameter choices" (Chomsky, 2009, p. 23). It is important to note that this idea of a developmental-genetic toolkit was developed independently of empirical research in developmental biology. Evo-devo theorizing is not representative of the research field at large. In fact, the conceptualization of evolutionary conservatism as an essentially *non-adaptive* process was deliberately set against the classical Darwinian notion of evolution by natural selection.[10]

To conclude, the cognitive model proposed by the generative theory is far less straightforward than is commonly conceived. It is not a clean analogy to the initial states and transitions to attained states of instinctive behaviors; at least not in Chomsky's *non-adaptationist* conceptualization of the language module. According to this notion, the transition from the initial to the attained state, too, is guided by endogenously determined processes of growth and maturation. The legitimacy of this argument is not taken from research on biological instincts. It is taken from a special line of theorizing in evolutionary developmental biology (evo-devo). The reason for this complication of the argument owes to the recognition that human languages rely on symbolic reference and, moreover, display unpredictable structural variation cross-linguistically. To bypass this problem, the P&P approach adopts a view of structural parameterization known as *switch box* or *toolkit* model. According to this view, the morphological organization of extant species rests on the different parameterizations of a universal Bauplan, a small set of invariant genetic principles, from which highly diverse phenotypes—from fruit fly to human being—are derived during early ontogenetic development via slight changes in the timing and switching on and off of regulatory genes. By analogy, the FL hypothesis, too, rests on assumptions about regulatory genetic development: there exist genetic switches for a human FL system that determine whether a human individual acquires an SVO or SOV language, a pro-drop or non-pro-drop language, a language with or without prepositions, and so on. However, this possibility of an ontogenetic switch box mechanism completely *redefines our notion of biological instinct* and assumes that different sets of attained states grow and mature in different linguistic environments.[11] This process of maturation is no longer based on the notion of learning by experience, a criticism also taken seriously by generative linguists themselves.

The Adaptationist Approach

Even among proponents of the generative (nativist) approach, the non-adaptationist version of the FL hypothesis has received much criticism. According to Pinker and Jackendoff (2005), for example, FL is precisely that: a complex and functionally *adaptive* system involving a great number of coevolved traits, some of which are language-specific (e.g., vocal imitation of speech, phonological structure, morphology, recursion,[12] and a mechanism for mapping sound onto meaning) and some of which belong to more general cognitive abilities that are either human-specific or shared with other animal species (e.g., theory of mind; spatial, causal and social reasoning). Hence, unlike Chomsky and others, proponents of the *adaptationist view* explain the structure of FL both in functional and evolutionary terms as this would apply to any other biological system. "Though we know few details about how the language instinct evolved, there is no reason to doubt that the principal explanation is the same as for any other complex instinct or organ, Darwin's theory of natural selection" (Pinker, 1994, p. 333). Even more explicit, Pinker and Bloom (1990) state that "the ability to use a natural language belongs more to the study of human biology than human culture; it is a topic like echolocation in bats or stereopsis in monkeys, not like writing or the wheel" (Pinker & Bloom, 1990, p. 707).

The adaptationist approach thereby constitutes the most faithful attempt to conceptualize the acquisition of grammar in true analogy to the development and structural organization of other biological systems. A major incentive for this reformulation has been the realization that the non-adaptationist FL hypothesis is ultimately driven by notions of optimal design and structural invariance (= symbolic computation) not compatible with notions of functionally adaptive design in the context of biological evolution (Johnson & Lappin, 1997, pp. 327, 330). Moreover, proponents of the adaptationist approach (e.g., Pinker & Bloom, 1990; Pinker, 1994; Stromswold, 1995; Jackendoff, 2002; Pinker & Jackendoff, 2005) do not share the skepticism adopted by non-adaptationist theorists (e.g., Chomsky, 1995, 2002a, 2002b; Piattelli-Palmarini & Uriagereka, 2004; Piattelli-Palmarini, 2009; Berwick & Chomsky, 2016). They propose that innate learning biases together with empiricist-style learning can adequately explain the acquisition of symbolic-yet-functional grammatical knowledge of individual languages.

The adaptationist argument is thus built on the assumption that children acquire functional knowledge of a language based on a complex set of innate learning biases for language-specific categories such as phonemic inventory, word structure, word-class membership, grammatical relations, inflectional morphology, and so on. If true, this acquisition process is indeed analogous to that of other complex abilities in humans and animals alike, such as, for instance, learning how to visually categorize objects into different colours, shapes, sizes, configurational patterns, motion events, and so on, based on innate perceptual biases in the initial state (e.g., see reviews by Logothetis & Sheinberg, 1996; Tarr & Cheng, 2003). If true, however, this acquisition process also requires the existence of reliable perceptual cues from the linguistic environment. To be of use to the child, such input has to be inherent to the acoustic/visual features of the presented speech stream—whether spoken or signed. If available, such universal markers of grammatical information could, in principle, "bootstrap" the child into selectively directing its attention to linguistically relevant units in a language (Höhle, 2009). On this basis, the child could learn, in a context-dependent manner, all the lexical and formal (parametric) particularities of a given mother tongue (Pinker, 1984, 1987; Landau & Gleitman, 1985; Gleitman, 1990; Jusczyk et al., 1992; Stromswold, 1995).

Experimental psychologists and linguists alike have taken up this claim to test its predictions in the lab. "The attentiveness on the part of newborns to the sound properties of native language utterances fits well with the view that sensitivity to cues in the acoustic input may be one of the important routes that infants take in discovering the set of grammatical relations that hold for utterances in their language" (Jusczyk et al., 1992, p. 289). "One possibility is that children have innate mechanisms that predispose them to perceive categorically such linguistic stimuli as phonemes, words, syntactic categories, and phrases, and that exposure to these types of linguistic stimuli facilitates the neural fine-tuning necessary for normal language acquisition" (Stromswold, 1995, p. 866). "Rather than being exclusively hardwired for speech or sound, the young of our species are initially hardwired to detect aspects of the patterning of *language*—specifically, aspects of its rhythmic and distributional regularities—corresponding to the syllabic and prosodic levels of natural language organization" (Petitto, 1999, p. 48).[13]

For at least well over two decades now, researchers have looked for such universal markers of grammatical information in the human speech code and a human-specific ability to detect them.[14] The findings of such research, however, and in particular the interpretations given to them, are highly ambiguous. On the one hand, such studies often highlight the fact that individual languages do possess conspicuous regularities in phonological makeup, prosodic structure and morphological marking, and, moreover, that infants can be shown to attend to such rhythmic features of human speech when familiarized with them under natural or experimental conditions (e.g., see studies conducted in psycholinguistic and neuroscientific labs headed by P. Jusczyk, P. Kuhl, A. Friederici or L.-A. Pettito). Yet, on the other hand, such studies literally never demonstrate that natural languages concur in the kind of physical input features (phonetic, prosodic, morphological, etc.) on which they happen to draw categorical distinctions in grammatical structure. As Stromswold (1995) incidentally highlights the dilemma: "One simple possibility is that every verb in every human language shares some readily accessible property, and children are innately predisposed to look for this property." But, "[u]nfortunately, there doesn't seem to be any such property" (p. 857). Likewise, Jusczyk and colleagues repeatedly point out that the correlation between acoustic markers and grammatical categories is notoriously "imperfect" even within individual languages (Jusczyk et al., 1992, p. 254). "None of the many available segmentation cues is foolproof" (Johnson & Jusczyk, 2001, p. 548) in the sense that none of the tested cues ever reliably correlate with only one syntactic feature in a language (e.g., phonetic cues with phonemic distinctions, prosodic cues with word entities, or intonational cues with clausal boundaries).

This notorious *unreliability of linguistic stimuli* is very unlike the physical reliability of perceptual cues in other biological systems (e.g., visual/auditory perception; spatial orientation). Here, individuals of the same species can, for specific domains of behavior, be shown to rely on exactly the same kinds of perceptual cues on the basis of which they learn to plan and direct their behavior under variable conditions. For example, despite external variability, organisms such as the barn owl can consistently rely on acoustic signals and the internal time differences they create to derive reliable estimations of spatial coordinates (see review by Knudsen, 1999). Moreover, due to the universal reliability of the physical laws operating on acoustic signals (e.g., their time of travel through space), the auditory system dedicated to sound localization has evolved anatomical and physiological prespecifications in the brain anticipating the processing of auditory cues resulting from such acoustic signals. These even include initial biases in behavior (e.g., characteristic head-body movements) facilitating the individual's primary interactions with the environment. Likewise, honeybees (see review by Menzel & Giurfa, 2006) can rely on a multitude of visual cues associated with the detection of potential food sources (e.g., ultraviolet light, configurational patterns) and spatial orientation (e.g., landmark contours, landmark arrangements, the azimuth of the sun). These cues can be identified as reliable visual features of both potential food sources (flowers) and navigable landscapes (open fields, meadows). Hence, again, one can motivate why individuals possess strong initial biases to attend to some as opposed to other aspects of the visual environment. They pick out those cues reliably associated with specific, ecologically relevant goals of behavior. "Such stimuli can be readily classified as the same because they involve a small number of easily measured physical parameters" (Hurford, 2007, p. 22).[15]

The linguistic environments surrounding small children, by contrast, do not yield such reliable input. The phonological, prosodic, morphological, and syntactic makeup of natural languages does not systematically correlate with grammatical structure in an analogous way such that *form* ("physical contour") could predict *meaning* ("functional value"). Instead, Chomsky (2002a) critically observes:

> Each expression [. . .] is an internal object consisting of two collections of information: phonetic and semantic. These collections are called "representations," phonetic and semantic representations, but there is no isomorphism holding between the representations and aspects of the environment. There is no pairing of internal symbol and thing represented, in any useful sense. (Chomsky, 2002a, p. 87)

The adaptationist approach, however, is built on precisely this assumption of *nonsymbolic* (isomorphic) representation. "'Bootstrapping' solutions [. . .] propose that a universal and innate linkage exists between the meanings of lexical items and the syntactic form classes to which they belong. This linkage is crucial in allowing children to match phonological strings with the appropriate meanings and functions in their native languages" (Danziger, 2008, p. 30).

To conclude, with its notion of a language organ in true analogy with other biological systems, the adaptationist approach has postulated no less than is required of it. Moreover, in its most influential formulations (e.g., Pinker, 1994; Jackendoff, 2002), the theory has properly taken into consideration many insights gained from cognitive research. It has properly acknowledged that the parametric fine-tuning is gained from continuous interaction with the environment (i.e., experience). Thus, even if born with clear language-specific biases, the child is still expected to learn lexical meanings in context (e.g., forming associations between /dog/ and 'dog,' /run/ and 'run'); it is still expected to learn the particular phonological structure of words irrespective of the innate ability to demarcate word boundaries; and it is still expected to learn the morphological realizations of grammatical categories particular to each language (e.g., formal tense, number, agreement marking). The child is assumed to proceed in very much the same way that, for example, a honeybee learns to associate specific instances of bilateral symmetry with individuated types of flower to acquire functional knowledge about reliable food sources.

Nevertheless, natural languages do not conform to any such universal linkage between meaning and syntax. There exist no universally reliable perceptual cues in the presented speech stream on the basis of which children could pick out phonemic differences in a language, word boundaries, the referential meanings of words (e.g., object-denoting versus action-denoting), inflectional markings (number, person, tense, etc.), or the relational functions attributed to phrasal entities in different argument positions (subject, object, etc.). Thus, so far, there is no evidence for the existence of universal linguistic markers (e.g., phonological, morphological, prosodic) that are indicative of such basic categories as tense, subject, noun, or verb. On the other hand, biologists do identify such reliable universal markers in nonlinguistic biological systems (e.g., physical location-, landscape-, flower-, face-markers). This finding suggests that the adaptationist reformulation of the FL theory, too, is inadequate in explaining the evolutionary foundation of natural human languages.

THE GENERAL INTELLIGENCE HYPOTHESIS

The functional (cultural) theory of language connects itself with a different line of research in cognitive science. This research sets the acquisition of language on a domain-general cognitive basis of learning. It thereby also sets up a fundamentally different hypothesis with regard to the origin and emergence of grammatical structures in natural language use. According to the theory, natural languages are of *cultural* not biological origin. Linguistic phenomena are the result of social traditions of usage and their passing on (transmission) across the generations. Consequently, there exists no initial cognitive state specific for the acquisition of the linguistic phenomena investigated by linguists. Linguistic competences in the individual are built on and emerge from domain-general cognitive abilities. These in turn are to a great extent shared with prelinguistic, nonhuman primates.[16] Some works that have been influential within this functionalist research tradition and that present interdisciplinary efforts between linguistics and the cognitive sciences are Savage-Rumbaugh (1986), Donald (1991), Deacon (1997), Savage-Rumbaugh et al. (1998, 2006), Tomasello (1999, 2003, 2008), Burling (2005), Evans and Levinson (2009), and Everett (2012, 2017).[17]

Intelligence or (henceforth) *general intelligence* is the term most frequently used by cognitive scientists to refer to the higher cognitive abilities of animals and humans that distinguish their behavior from more primitive, context-specific interactions with the environment (Darwin, 2004 [1879]; Thorpe, 1963; Byrne, 1995; Deary, 2001; Russon & Begun, 2004; Geary, 2005; Reader et al., 2011; Call, 2013). The term *general* implicitly stands in opposition with terms such as *instinctive*, *reflexive* or *domain-specific* and thereby carves out a categorical distinction between decision-making based on experience, voluntary control, and reasoning as opposed to behaviors based on no prior experience, involuntary responses, and no reasoning. Yet, given that animal and human behaviors are extremely diverse and differ in complexity, scholars agree that the capacity of general intelligence has evolved by degrees and is therefore best characterized along a scale or continuum of increasing complexity rather than attributed as an all-or-none ingredient of human or nonhuman behavior. "Their technicalities apart, all such studies confirm the common 'folk' assumption that non-human animals, especially mammals and birds, while falling short of humans in many of their mental abilities, nevertheless learn, remember, and plan" (Hurford, 2007, p. 7). It is therefore "useful to talk in terms of an evolutionary succession: proto-concepts → pre-linguistic concepts → linguistic concepts" (Hurford, 2007, p. 12). "Some of the semantic properties that we see in modern human languages can be traced back to these pre-linguistic foundations" (Hurford, 2007, p. 8). As is indicated here, general intelligence is associated with the ability to acquire *conceptual knowledge* about the world and to act on this basis in a flexible and context-dependent manner. On the scale offered by Hurford, ranging from proto-concepts to linguistic concepts, *prelinguistic concepts* mark the critical stage in cognitive evolution that ultimately also launched the evolution of language within the hominin (Homo) lineage.

But what are the criteria for domain-general cognition and the entertainment of prelinguistic concepts? And at what stage along the continuum do these general learning abilities enable subjects to initiate and engage in linguistic behavior? Next, we tackle each question in turn.

The Development of Prelinguistic Concepts

The cognitive sciences have brought fourth an immense wealth of empirical evidence, which shows that nonhuman animals are endowed with more powerful learning capacities, be these domain-specific or general, than was originally anticipated by the nativist turn in the mid-twentieth century. "Many animals other than humans, especially mammals and birds, possess well developed knowledge-of-the world (declarative memory) systems, and are capable of acquiring vast amounts of flexibly expressible information" (Tulving & Markowitsch, 1998, p. 202). Such flexibility already begins to show in very specific domains of behavior. The recognition of potential predators, prey, important food sources, social others (faces) or the coordination of flight are all examples of responses to the environment for which innate dispositions exist. Yet, even here the fine-tuning of such innate reflexes depends on experience, that is, repeated encounters with the environment.[18] "Even reflex actions involve some kind of generalization over stimuli. No two stimuli are ever exactly the same. The various slightly different stimuli prompting reflex actions are at the bottom of a scale of complexity in classifying things or events" (Hurford, 2007, p. 22). On Hurford's scale, these generalizations qualify as "proto-concepts." They are based on generalization and experience; however, they still lack the voluntary control and domain-general aspect characteristic of more advanced behaviors.

Prelinguistic concepts are first recognized in areas of problem-solving where individuals begin to generalize across sets of stimuli from various sensory modalities. Moreover, they allow swift and context-dependent reorganizations of behavior. The typical experimental paradigm is one where individuals are trained to associate complex stimuli with the presence or absence of desirable food sources. During testing, they are exposed to novel stimuli that still resemble the ones from training, yet only by extension, that is, generalization.[19] In this way, various nontrivial categorization abilities have been tested in animals with positive results. For instance, chimpanzees can learn to sort novel items based on the distinction between edible and nonedible objects (Savage-Rumbaugh, 1986). Pigeons can learn to distinguish pictures containing people from pictures not containing people (Herrnstein et al., 1976). Even more impressively, they can learn to distinguish different schools of art (paintings in the tradition of Monet versus Picasso) (Watanabe et al., 1995). Evidence of the latter kind is considered more significant as the tested stimuli do not easily resemble the kind of stimuli that are deemed relevant under natural conditions. Often, the literature only focuses on vertebrate (mammal and bird) cognition to demonstrate higher learning abilities. However, in fact, many studies also show such cognitive plasticity in invertebrates. Thus, the German neurobiologist Randolf Menzel and his colleagues have demonstrated an astonishing degree of flexibility even in such a small animal as the honeybee. Honeybees can learn to distinguish between different color paradigms (Giurfa, 2004), black-and-white patterns (Giurfa et al., 1999), and even recognize bilateral symmetry on a par with the requirements of human IQ tests (Giurfa et al., 1996). Moreover, they can solve matching-to-sample tasks involving the flexible transfer from the olfactory to the visual modality (Giurfa et al., 2001).

> As in larger brains, two basic neural architectural principles of many insect brains are found: specialized neuropiles, which refer to specific sensory domains, and higher-order integration centers, in which information pertaining to these different domains converges and is integrated, thus allowing cross-talk and information transfer. In this sense, both modularity and central integration seem to be basic building principles adopted by different nervous systems to provide flexible solutions to a changing environment. (Menzel & Giurfa, 2006, p. 37)

Hence, prelinguistic concepts require "abstract" or "structural" representations derived from central integration (Call, 2013, p. 12). Abstract or structural representations still rely on external input; however, perceptually more complex configurational properties of things come into focus. Objects are not only perceived in terms of simple features like color, smell, or shape but also in terms of texture, weight, consistency and also in relative terms in comparison with same- or other-kind properties of other objects. These more abstract, relational perceptions allow more sophisticated inferences about the world. The acquisition of such knowledge, however, also requires more interaction and multiple, cross-modally combined perceptions of the world. Properties like weight or consistency cannot be perceived unless the things are being interacted with in myriad ways, for example, by lifting, throwing, waving, hitting, and touching. That is, complex object manipulation is a necessary process by which such knowledge is gathered. Other than a predisposition to explore certain items in the world, there exists no prior knowledge of the complex information thus acquired (Byrne & Russon, 1998; Call, 2013). Moreover, the acquisition process also requires a larger "global neuronal workspace" in the brain to process such different channels of information in relation to one another (Dehaene, 2014, chapter 5). Hence, the neural capacity, and the proportion of brain areas associated with "cross-talk and information transfer" (Menzel & Giurfa, 2006, p. 37) is considered an important indicator of general intelligence. In this respect, nonhuman primates and of course humans outperform most other nonhuman animal species (Herculano-Houzel, 2016).

Another important aspect characterizing prelinguistic conceptual thinking—and thus arguably the kind of general intelligence ultimately also required for the acquisition of linguistic behavior—is the ability to apply abstract knowledge about the world in a context-dependent, flexible manner. "Having a concept involves some ability to 'stand outside' one's immediate behavior, possibly to choose one's next move, or evaluate alternative possibilities" (Hurford, 2007, p. 29). Spatial navigation experiments as well as food retrieval studies provide some telling examples. In spatial studies, individuals are trained in a confined space, for example, a maze (rats) or a field (bees), which contains conspicuous landmarks that are known to play a role in spatial orientation of these animals. During testing, they are exposed to novel arrangements, which require that the individual deviate from some previously acquired routine. For example, honeybees can successfully navigate their way home after being (unexpectedly) dislocated in relation to the beehive (Menzel et al., 2000, 2005). The results reveal that these animals do not only consult fixed flight routines in relation to specific landmarks. Instead, their behavior gives evidence to complex spatial memories (cognitive maps), which they recall in a context-dependent and goal-oriented manner. They can fly detours if they choose to, just as they can fly the shortest way back to the beehive from a novel, hitherto unknown position. In food retrieval studies, in turn, the focus has been on American scrub jays' ability to retrieve hidden food items based on previous exposure to various episodes (Clayton & Dickinson, 1998; Clayton et al., 2001). For instance, subjects witness how food items with different durability are buried in different places; time passes and individuals are then given free access to retrieve the food. The birds in most cases correctly choose those food items that are still good, not those which are bound to have perished. Likewise, they recall when some food item has been retrieved before and thus avoid trial and error. Such abilities for *episodic memory* and to *reason* about possible outcomes of behavior are all essential for the attribution of general intelligence.

Naturally, studies on model organisms like insects, rodents, or birds do not draw the full analogy to human cognition and behavior. These nonhuman animals are chosen for experimental study precisely because their brains are smaller than the human brain and because the behavior is less complex: "they show, in embryonic form, characteristics that are sometimes thought to be unique to humans" (Hurford, 2007, p. 8). In particular, they show that *intentionality* plays, already at this stage of prelinguistic interaction with the world, a crucial role during decision-making and the production of novel behaviors (Menzel, 2017).[20] Individuals pursue goals and rely on their various mental capacities of immediate perception, memory retrieval, comparison, inference, and future planning as internal processes on which they rely to negotiate their next moves. At a minimum, such intentionality requires that the individuals can entertain various *attitudes* or mental states toward the world. A honeybee that perceives a spatial landmark can adopt different attitudes depending on its goal orientation. It can recognize it as a positive cue for being on the right path toward a desirable patch of flowers. But, given that it has just been dislocated during its return flight to the beehive, the bee can also recognize it as a negative cue for being on the wrong path toward home. The landmark will thus assume different meanings for the individual depending on its current goal. There is nothing reflex-like about such context-dependent information processing.[21]

In humans, such intentionality is much extended. They possess far more powerful memory capacities and can infer a lot more information about current states of affairs based on general world knowledge and based on the knowledge about things currently out of view. Hence, their ability to recall objects, scenes, events and to include various experiences into their current modes of action is a very conspicuous property of human behavior (Donald, 1991; Suddendorf & Corballis, 1997). It allows them to act under what is commonly referred to as *displaced conditions*. Humans can plan far ahead and do things in the present that prepare them for much later. In fact, their habitual mode of living is to perform actions that are detached from what is going on in their immediate surroundings.

> The capacity to know something about an object, even when "it isn't there," is a first step along the road to the impressive characteristic of human languages: their capacity for displaced reference. "Displacement" was noted by Hockett as one of his list of "design features of language" (Hockett 1960; Hockett &

> Altmann 1968). We can talk about things very distant from our senses, such as the far side of the Moon and last year's potato crop. An animal that cannot represent an object out of its view cannot begin to work with displaced reference. Humans' ability to create labyrinthine mental models of spaces, and the movement of objects within them, without direct appeal to perception, exceeds the capacity of any non-human. As the psychologist's visible and invisible displacement tasks is a matter of amount of mental computation involved, the impressive abilities of humans in this regard may be seen as quantitative, rather than a qualitative advance over non-human animals. (Hurford, 2007, pp. 40–41)

To conclude, the term *general intelligence* as used in cognitive science summarizes various aspects of information processing that are constitutive of behavior that is increasingly brought under the volitional control of the planning and acting individual. The information is increasingly complex in the sense of multimodal integration of information and acquired through complex interactions with the environment. It is therefore referred to as structural or abstract rather than perceptual or sensory. Furthermore, it is acquired through various modes of learning that include different modes of interaction with the world, from explorative to observational to goal-directed (Thorpe, 1963; Call, 2013). Finally, the flexible, context-dependent retrieval of such structural knowledge is given a new platform of learning through its anchorage in context-dependent, intentional frames of mind (Malle et al., 2001; Menzel, 2017). As the negotiation of information is less guided through direct external reinforcement and more dependent on internal modeling and recombination of information, both planning and execution of action become more detached from the world outside.

The Development of Linguistic Concepts

In the late 1980s, Savage-Rumbaugh and colleagues carried out an unprecedented study involving 8-year-old Kanzi, a male bonobo (*Pan paniscus*), and 2-year-old Alia, a female human infant (*Homo sapiens*). The study (Savage-Rumbaugh et al., 1993) compared their abilities to understand spoken English sentences and to respond to them with concrete actions. The challenges put before Kanzi's language comprehension skills in this study were anything but trivial. First, Kanzi was asked to decode auditory verbal inputs with an underlying phonemic structure unique to language; this he was required to do without himself being able to encode such utterances by means of vocal articulation. Second, he was confronted with spoken test sentences that he had not encountered before under everyday conditions (e.g., "Feed your ball some tomato," "Can you put the ball on the pine needles?"). They therefore required him to comprehend relational meanings beyond mere word recognition. And third, Kanzi (just like Alia) was exposed to these sentences under firmly decontextualized and partly displaced conditions: for each test sentence, both subjects were presented with whole arrays of objects in front of them (e.g., Kanzi sitting on the floor and facing things like bowls, containers, cans, a knife, toothbrush, can-opener, ball, pine needles). Furthermore, they were often required to carry out actions involving more distant or even out-of-sight parts of the testing area (e.g., refrigerator, microwave, adjacent rooms). Considering the overall success rate of both subjects (fairly similar to one another, albeit slightly higher for Kanzi than for Alia), Savage-Rumbaugh et al. (1993) concluded positively that the nonhuman primate, Kanzi, had acquired some true linguistic competence. "The lack of contingent reward, the novel nature of the requests, and the absence of previous training to perform these specific requests, and the unique nature of each trial countermand simple explanations that depend on the conditioning of responses independently of semantic and syntactic comprehension" (Savage-Rumbaugh et al., 1993, p. 98).

This acquisition of linguistic concepts in a normally prelinguistic animal (*Pan paniscus*) is explained by two lines of research. One of these rests on the evidence already given in the previous section, namely that nonhuman animals are capable of entertaining complex trains of thought involving domain-general memory formation and retrieval, inferencing and the planning of action under displaced conditions. This capability reaches different degrees of complexity in different animal taxa and, arguably, approximates that of humans in those taxa most closely related to them. Hence, nonhuman primates in many ways closely resemble humans in their perceptions of the world and their general capacities to comprehend and memorize complex real-world scenarios. This explanation of primate cognition rests on the same assumptions of general intelligence as introduced for nonhuman animals more generally, and thus (conceptually) requires no further elaboration. We need only point out that what Kanzi was able to achieve in the study rests on very sophisticated abilities of object and event representation, including a wide array of object–object relations under displaced conditions. Thus, prelinguistic understanding of the world plays a vital role in linguistic comprehension (and by implication also production) and is not specific to the language domain.

The second line of research is more specific as to what constitutes linguistic behavior as opposed to nonlinguistic behavior. It seeks to explain how linguistic communication rests on special forms of prosocial, collaborative interaction. Thus, as was outlined earlier, Kanzi had learnt to associate arbitrary verbal signals with real-world actions and he understood their attention-guiding function in relation to himself. He had acquired a basic form of *symbolic*, *referential competence*. Conversely, he could also produce simple communicative speech acts himself. He could use portable, visual lexigram symbols to communicate his intentions to his human caretakers (Savage-Rumbaugh et al., 2006). For instance, he produced the lexigram combination "Car trailer" when he was sitting in a car and wanted to be driven in the car to a nearby trailer. Likewise, he produced "Grouproom Matata," where it appeared that he wanted his mother, Matata, do be brought to him into the grouproom instead of going to her. This ability to engage in *triadic interactions* with the world (Hubley & Trevarthen, 1979; Tomasello, 1999), involving a subject, an object and a social conspecific, had been particularly nurtured in Kanzi from his earliest infancy. Savage-Rumbaugh and colleagues had deliberately moved away from the classical training method of instructing subjects in long and repeated trials to memorize lists of word items (lexigrams) out of context (Savage-Rumbaugh et al., 2006). Instead they had focused on practicing daily sociocultural routines among peers and human caretakers (playing games together, preparing and sharing food together, going outdoors together, etc.). As the case of Kanzi first demonstrated, subjects reared under such *socially scaffolded, prosocial conditions* begin to learn and use symbols quite spontaneously, very much like children do during their first two years of life.[22]

The studies by Savage-Rumbaugh and colleagues have thereby demonstrated the importance of prosocial interactions for the establishment of common ground and the basic foundation on which intentional communication rests. Linguistic means are only "symbolic" in their relation to the things denoted by them (i.e., real-world referents). However, in usage, they are grounded by intentional interactions with the world. They are not directed at the things they denote; they are directed at the intentional mental states of other individuals who share a common ground of real-world experience with the language user. Hence, the functional anchorage of linguistic means is not at the interface between the subject and the physical world; it is between the subject and other subjects, and their various attitudes toward the world.

As subsequent research, most notably by Michael Tomasello and colleagues, has established more clearly, nonhuman primates are not normally intrinsically motivated or reared in a social environment where collaborative, prosocial interactions are predominant. Rather, nonhuman primates either interact intentionally with their physical environment, or they interact intentionally with their social environment. They prefer dyadic interactions (Zlatev et al., 2008a). But they do not often combine both domains to yield triadic, collaborative actions to perform tasks together (Tomasello et al., 2005; Moll & Tomasello, 2007; Tomasello, 2009). It is at precisely this interface, however, that intentionally directed communication sets in, even at a prelinguistic stage (Warneken et al., 2006; Warneken & Tomasello, 2007; Tomasello, 2008, chapter 4). Collaboration requires mutual coordination, and mutual coordination requires adjusting and occasionally guiding and readjusting each other's attentions. It is here that the use of communicative and ultimately linguistic means unfolds its true adaptive functionality.

Yet, arguably, this line of research does not account for *the compositional and conceptual aspect* of linguistic behavior. This, however, is the characteristic most in need of an explanation in terms of language evolution because there exists no direct precursor in prelinguistic behavior. Animals normally do not speak. Hence, what is so impressive about Kanzi's achievement is his ability to derive meaning from combinations of linguistic symbols and to interpret their meaning on a syntactic, relational level. In order to respond with actions like "putting a tomato on a ball" or "fetching a toy from another room," he had to infer meaning from word order and phrasal units specifying agents, patients, themes, locations, goals, sources, and so on. Similarly, he had to gain some basic knowledge of generalizations (i.e., patterns of transfer) associated with linguistic symbols. In comparison to prelinguistic concepts, linguistic concepts are associated with much broader generalizations. Even the simplest linguistic items like words for *red*, *fork*, *teddy bear* or *bowl* delineate categories across real-world objects not easily aligned with primary percepts of the environment. The English word *red*, for example, covers an extremely wide array of visual stimuli. Subjects can have many different impressions of RED; yet, linguistically, they are all treated the same, namely as *red*. Hence, linguistic symbols are recruited very freely and their link to primary perceptions of the environment is opaque and only schematic. Apparently, Kanzi was still able to infer such generalizations based on some prelinguistic intelligence. This ability is still in need of further explanation. So far, the functionalist research tradition does not offer a satisfying solution to this quest for an answer.

FINDING THE MISSING LINK

The whole point of Chomsky's initial postulation of a language module and his insistence that it relies on symbolic, not isomorphic, computation of real-world experience was his recognition of the fact that "the connection to the outside world is extremely weak" (Chomsky, 2002b, p. 148). Biological systems, because they involve neural plasticity, are controlled by the outside world. A honeybee that has lost its path during navigation has to readjust its active mental state or working memory to the physical surroundings. It cannot imagine a different landscape and pretend that it has found a way back home unless this route correlates with the physical reality outside. Human spatial navigation has to meet the same requirements. As opposed to such nonlinguistic biological systems, "language doesn't have to meet that condition, or it has to meet it in an extremely weak way. You have to be able to talk about the world, somehow, but there's any number of ways of doing that" (Chomsky, 2002b, pp. 147–48). Hence, the adaptationist approach to the faculty of language hypothesis fails. However, the non-adaptationist approach also fails, because it doesn't meet the requirements of functional-structural adaptation during the transition from the initial to the attained state.

The general intelligence approach, by contrast, offers a new research avenue in order to explain the emergence of human linguistic behavior. It relies on a model of learning that acknowledges the existence of "specialized neuropiles" dedicated to the processing of primary sensory inputs from the environment. Yet, it also recognizes the existence of 'higher-order integration centers in the brain which allow the flexible cross-talk and transfer of information between different sensory modalities' (Menzel & Giurfa, 2006). Such higher-order information processing, sometimes called abstract or structural representations of the environment, allows an individual to act on the retrieval of memory in such a way that intentions can be operated on; as a result, its behavior becomes more flexible and context-dependent. The neuronal basis of such "global workspaces" (Dehaene, 2014) are known as "cognitive maps" that provide complex, holistic representations of the environment (e.g., landscapes, inanimate objects, animate beings, episodic memories; see Tolman, 1948; Thorpe, 1963; Carew, 2000; Menzel & Giurfa, 2006). On this basis, animals plan and readjust their behaviors in ongoing activities at the level of the global work space, besides also relying on conditioned responses as well as fixed routines acquired under stable conditions of the environment.

The general intelligence hypothesis thus goes a long way to explain several undeniable characteristics of the human ability to speak. In particular, ape language studies have shown that nonhuman primates already possess complex understandings of real-world scenarios enough to properly decode complex linguistic utterances. Moreover, they show that a culture of collaborative interaction greatly facilitates the acquisition of linguistic symbols (Savage-Rumbaugh et al., 1998, 2006). At the same time, comparative research provides evidence that human infants are intrinsically more motivated to engage in triadic interactions with their conspecifics. Given that nonhuman primates do not naturally communicate by linguistic means, it is a plausible assumption that the further extension of general intelligence to modes of "shared intentionality" formed an important starting point for the eventual arrival of natural languages (Tomasello, 1999, 2008; Tomasello et al., 2005).

This leaves us with the question of how the notion of "general intelligence" can also explain the compositional character of linguistic means (constructions) and how their underlying concepts relate to usage and real-world experience. In other words, we still haven't explained *the cultural origin of natural languages* and how general intelligence still forms the necessary and sufficient substrate of this development. The next chapter offers an answer to this question. It argues that the continuum from proto-concepts (instincts) to prelinguistic concepts (general learning abilities) to linguistic concepts (Hurford's 2007 lingo) must be complemented by one further link between prelinguistic and linguistic concepts. This link is referred to as *tool intelligence*. The tool notion is useful because it has already figured in linguistic theorizing. However, so far, it has not been synchronized with the debate on "language in the light of biological evolution." The current work offers an attempt to bridge this gap.

NOTES

1. Inter alia, Pinker (1994), Donald (1991), Bickerton (1996), Deacon (1997), Savage-Rumbaugh et al. (1998), Tomasello (1999, 2008), Chomsky (2000), Belletti & Rizzi (2002), Jackendoff (2002), Love (2004), Burling (2005), Hurford (2007, 2012), Cowley (2011a), Everett (2012, 2017), Berwick & Chomsky (2016).

2. The intellectual background to the biolinguistic perspective is outlined in Chomsky (2009, pp. 13–19).

3. Chomsky draws on quotations from C. R. Gallistel in this passage. Though initially reserved to the faculty of language, this *modular view of learning* was subsequently also postulated for other cognitive systems (Samet & Zaitchik, 2017 [2012], chapter 1.2, pp. 11–16).

4. For a more detailed consideration of other models of formal grammar developed at the time, see Johnson & Lappin (1997). As these authors report, Chomsky's version of the P&P approach (Chomsky, 1980, 1981) was preceded by other "constraint-based approaches to grammar" such as Arc Pair Grammar (Johnson & Postal, 1980), and developed at around the same time as "unification-based frameworks" such as Lexical Functional Grammar (Kaplan & Bresnan, 1982), Functional Unification Grammar (Kay, 1983), or Head-Driven Phrase Structure Grammar (Pollard & Sag, 1994). As all of these approaches have in common that they are varieties of generative grammar and since Chomsky's work in particular has shown a continuous interest in the cognitive science debate, we do not consider these other models in our debate. However, problems arising for Chomsky's generative nativism should in principle, in one form or another, also apply to these other approaches.

5. E.g., see works by Massimo Piattelli-Palmarini and colleagues (e.g., Piattelli-Palmarini & Uriagereka, 2004, 2008; Boeckx & Piattelli-Palmarini, 2005; Piattelli-Palmarini, 2009). More generally, the *modular view of learning* can be associated with works by C. R. Gallistel (1990, 1999), D. W. Thompson (1992 [1917]), and Jerry Fodor (1983).

6. The neurobiologist Randolf Menzel reminds me (p.c.) that the notion of "learning (by experience)" involves many more types of learning than "associative learning." For instance, there is also explorative learning, latent learning, or observational learning. I agree and also take note of this fact in the discussion further below (subsection "The Development of Prelinguistic Concepts"). However, I do not agree

that this lack of specification in the present discussion opens up a false dichotomy between "learning by experience," on the one hand, and "instinct," on the other (as Menzel maintains). I believe that close analysis of the generative literature (in Chomsky's tradition) really leads to the insight that "learning by experience" cannot be either associative or explorative or any other type of learning; it can only be an endogenously predetermined process of growth and maturation that happens *independently* of information from the environment. It may well be that the modern cognitive sciences deem even such notions of "biological instinct" as outdated, but the task of the current section is to explain what premises the generative theory is based on, not to discuss the present state of the art in the cognitive sciences at large. According to my argument, the present state of the art is most faithfully recapitulated in functionalist accounts of language learning, which I turn to in the second part of this chapter (section "The General Intelligence Hypothesis").

7. See some of the classical studies from ethology (Lorenz & Tinbergen, 1938; Tinbergen, 1951; Lorenz, 1981), neuroethology (e.g., Ewert, 1974, 2004, on prey detection in amphibians; Chan et al., 1998, and Tammero & Dickinson, 2002, on flight coordination in flies; Wyttenbach et al., 1996, on sound perception in crickets), and human ethology (e.g., Goren et al., 1975, Johnson et al., 1991, and Tarr & Cheng, 2003, on face recognition in human infants). For an overview, see chapters in Carew (2000).

8. For example, see Suboski (1992) for a review of learning abilities in amphibians, in the context of prey-catching and avoidance behaviors. See Menzel & Girufa (2006, pp. 31–33) for a review on honeybee cognition and behavior, especially the discussion of selective attention and visual stimulus categorization. See Logothetis & Sheinberg (1996) and Tarr & Cheng (2003) for a review on visual object recognition in vertebrates. See Carew (2000) for overviews on various studies on model organisms in behavioral neurobiology.

9. Note that we draw a distinction here between studies in developmental biology and *evo-devo* theories. The term *developmental-genetic toolkit* has been coined within the relatively young field of evolutionary developmental biology (*evo-devo*), which has emerged through interdisciplinary efforts in developmental biology, molecular genetics and evolutionary biology. Especially for Chomsky, important figures (founding fathers) in this line of theorizing are Stephen Jay Gould and Richard Lewontin. For more detailed discussions of evo-devo research, see Goodman & Coughlin (2000) and Müller & Newman (2005). For an in-depth account of the developmental-genetic toolkit approach, see Gerhart & Kirschner's (2007) theory of facilitated variation.

10. Relevant works, which support Chomsky's non-adaptationist account of language evolution, are Gould (1977), Gould & Lewontin (1979), and, more recently, Gerhart & Kirschner (2007).

11. For a similar criticism, see Christiansen & Chater (2004, p. 492).

12. *Recursion*, or *Merge*, is the term that was introduced by Chomsky (1995) in his latest reformulation of the non-adaptationist FL hypothesis—the *minimalist program.* In an influential paper by Chomsky and colleagues, Hauser et al. (2002), recursion is considered the uniquely human capacity to produce a potentially infinite number of discrete expressions out of a finite set of elements (Hauser et al., 2002, p. 1571). This part

of the language faculty is called FLN (= faculty of language in the narrow sense), which is kept strictly separate from other processing capacities of the human brain. While other subsystems of FL are considered to be evolutionary adaptations for communication, FLN is not. This property makes FLN unlike any other mechanism in the human brain: it is an invariant computational system which by itself is not endowed with any specific function, only abstract rules. Regarding its recruitment as a communicative system, Hauser et al. speculate that it might have been "exapted" from other cognitive abilities such social reasoning, tool-making, or (spatial) navigation (Hauser et al., 2002, pp. 1570–72, p. 1578). The lack of functional specificity and invariance of FLN is of special importance. It implies that evolution by natural selection could not have shaped such a recursive mechanism. Rather, "certain specific aspects of the faculty of language are 'spandrels'—by-products of pre-existing constraints rather than end products of a history of natural selection" (Hauser et al., 2002, p. 1574). The notion of evolutionary "spandrels" is taken from Gould and Lewontin (1979). Along similar lines, Chomsky (2009) argues that the gradual process of evolution cannot explain the emergence of "unbounded Merge" (i.e., recursion). Instead, he considers it more plausible that "some small mutation" occurred at some stage in recent human history (some 100,000 years ago), leading to an abrupt reorganization of the human brain, and incidentally creating the recursive capacity (Chomsky, 2009, p. 29; see also Chomsky, 2002b, p. 149).

13. For further statements along the same lines, see e.g., Friederici (2002, p. 83), Kuhl (2004, p. 831), Mueller et al. (2012, p. 15953), Gervain & Werker (2013, p. 1). The research tradition is still well and alive.

14. E.g., see studies by Werker & Tees (1984), Cheour et al. (1998), Maye et al. (2002), Mahmoudzadeh et al. (2013), Partanen et al. (2013) with a focus on speech sound perception; see studies by Saffran et al. (1996a, 1996b), McQueen (1998), Mattys et al. (1999), Johnson & Jusczyk (2001), Friedrich & Friederici (2005) with a focus on word segmentation; see studies by Klatt (1976), Bolinger (1978), Jusczyk et al. (1992), Fisher & Tokura (1996), Soderstrom et al. (2003, 2005), Gómez & Maye (2005), Seidl (2007), Friederici et al. (2011), Wellmann et al. (2012), Mueller et al. (2012) with a focus on grammatical units beyond the mere word level. The interested reader is referred to Cowie (2017 [2008], chapter 3.3, pp. 54–64) for a concise overview and criticism of pro-nativist language acquisition studies; to Jusczyk (2000 [1997]) and Werker & Tees (2005) for reviews on experimental research conducted in the field of cognitive psychology; and to Friederici (2005) and Kuhl (2004, 2010) for reviews on research in the relatively young field of cognitive neuroscience.

15. Along the same lines, Chomsky observes: "So there's some innate system of object recognition: infants can identify object constancies; they know things don't go through barriers etc. But that system, whatever it is, has to be attuned to the outside world; if you had a system that had objects going through barriers and so on, you couldn't get along in the world. So that system is sort of controlled by the outside world" (Chomsky, 2002b, p. 147).

16. In this functionalist view, humans are not without species-specific endowments regarding, for example, the production of articulated speech or the speed and quantity of learning. However, the underlying genetic dispositions regarding such abilities are considered no more special than, say, the ability in some species of songbirds or

parrots to produce complex vocalizations (including the imitation of human speech), or the ability in some animal species to engage in manipulative behaviors (e.g., stick use in New-Caledonian crows; nut-cracking in wild chimpanzees). In more cognitive terms, the species-unique endowment of humans is often said to be linked to an increase in brain size and/or neural capacity, which has the general effect of enabling humans to learn more swiftly and to store and retrieve greater amounts of information while planning and carrying out goal-oriented actions (Gibson et al., 2001; Read, 2008; Dehaene, 2014; Herculano-Houzel, 2016).

17. Among such works of the functionalist tradition, some have also been explicitly concerned with criticisms of the generative theory of language. Among them are works by Savage-Rumbaugh et al. (1993, 1998), Tomasello (1995, 1998, 2003), Christiansen & Chater (2008), Evans & Levinson (2009), and Everett (2012, 2016, 2017). Within linguistics (i.e., not so much at the linguistics/cognitive science interface), the typologist Martin Haspelmath has for many years been a very forceful critic of the generative theory (e.g., see Haspelmath, 2000, 2007a).

18. E.g., see reviews by Logothetis & Sheinberg (1996), Tarr & Cheng (2003), Knudsen (1999), Sharp (1999), Menzel & Giurfa (2006), each focusing on different model organisms. For an overview, see Carew (2000).

19. In some follow-up experiments, the conditions are sometimes reversed so that a given association between a stimulus and a reward is directly opposite to what it was before. In such tests, it is measured how quickly individuals reverse the association; whether they have to relearn it entirely or can swiftly reverse it, showing that they abstracted a more general principle. For an overview of such experimental paradigms, see Menzel (2013).

20. In the philosophical tradition of Brentano (1973 [1874]) and Searle (1983), *intentionality* refers to the property of all mental states as being directed toward something, for example, objects, knowledge, beliefs. In the modern understanding among scholars focusing on human behavior, it refers to the property of actions that are done on purpose or intentionally (Malle et al., 2001, p. 3). Important attributes of such behavior and the mental states underlying it are the possibility to entertain different attitudes toward things (e.g., knowing, believing, desiring, imagining), as well as a certain degree of awareness and the ability to reason about different possible outcomes of behavior. Hence, intentionality (also Intentionality with a capital "I," see Hurford, 2007, p. 6) implies aboutness in the sense that individuals entertain different attitudes toward things/objects in the world. Such objects can be concrete and perceptible but also imaginary. Intentionality must be distinguished from *intention*. Intentionality is a quality of actions, while intentions describe the mental states that represent such actions (Malle et al., 2001, p. 3). The attribution of intentionality to human and nonhuman behavior thus depends on how broadly one defines aboutness and the degrees of awareness and reasoning capabilities underlying flexible behavior.

21. Similarly, the scrub jay who is given free access to a patch of hidden food does not treat each spot as though it contained desirable food items. It entertains differential attitudes toward each spot based on the time that has elapsed since previous exposure and based on the various episodes memorized to have taken place on the same turf.

22. Kanzi had come to the Language Research Center, Georgia, in 1980 together with his adopted mother Matata, who had been captured from the wild. From 6 months of age until the age of about 2 years, Kanzi had never been included in any of the explicit training sessions given to Matata, who proved to be a poor learner, and it was not until he was first separated from his mother for a few months that his human caretakers began to realize that he had already acquired a level of symbolic/linguistic competence outperforming that of his hard-trained adult peers (Savage-Rumbaugh et al., 2006, p. 529).

Chapter 3

The Tool Hypothesis

It is not uncommon to refer to language as a *tool* for communication. In fact, the tool metaphor has figured throughout the linguistic literature for at least a century. Perhaps, most famously, the German psychologist Karl Bühler chose the term *organon* (Greek: 'instrument, tool') to explain the communicative functions of language (Bühler, 1934). Likewise, the Russian psychologist Lev Vygotsky drew on the tool concept to explain the ontological development of higher cognitive faculties in humans and the internalization of speech. "The invention and use of signs as auxiliary means of solving a given psychological problem," he writes "is analogous to the invention and use of tools [. . .]. The sign acts as an instrument of psychological activity in a manner analogous to the role of a tool in labor" (Vygotsky, 1978, p. 52).[1] Toward the end of the twentieth century, especially linguists with a background in social and cultural anthropology kept this analogy alive.

> The tool has an impact on raw material [. . .]. Divorced from the context of production, the tool reverts to its original condition as an inert object [. . .]. The parallel [. . .] extends to the use of words in dialogue. In themselves, words have no meaning, they merely afford the possibility of meaning. Only by their being harnessed to the intention of a speaker, in the context of a verbal exchange, do they come to have a meaningful impact upon the listener. (Ingold, 1993, p. 440)

Goody (1997), another anthropologist, focuses on the relationship between social intelligence and language, and again highlights the tool functions performed by language in ongoing discourse: "language-in-use is a powerful *tool* for the co-operative construction of meaning; language-in-use is a new sort of *tool* for co-operation between individuals; and language is a *tool* for constructing socio-cultural devices for co-operation among members of a community" (Goody 1997, p. 366; emphasis added). Works in the sociolinguistic tradition reiterate such claims and emphasize the various "instrumental functions" of language in public and private discourse (Duranti, 2011).

More recently, the linguist Daniel Everett, known for his research on the Amazonian language Pirahã, dedicates a whole monograph to *Language: The Cultural Tool.* Cultural tools such as language, mathematics, music, and other material cultures, Everett explains, must be learned. "The dividing line rests on whether a tool is learned or just grows. Arms are not learned, they just grow. So they are biological tools. Numbers do not just grow, they are learned. So they are cultural tools. We do not learn to hear, but we learn to make music. So music is a cultural tool, but hearing is a biological tool" (Everett, 2012, p. 29). Everett therefore argues that the acquisition of language, like the acquisition of other cultural tools, is based on general intelligence:

> One could just as easily say that there is no innate linguistic capacity but that the human mind instead simply is a good problem-solver and uses some tools for a variety of purposes. We could hypothesize that in different societies the language problem is solved by similar, but non-identical methods that are not unique to language but are implicated in a range of different skills. (Everett, 2012, p. 89)

Hence, the tool metaphor is not new. For at least a century it has been an important point of departure for functionalist explanations of language evolution, language use, and language development. Nevertheless, despite its presence, it is not a theoretical concept with much impact in the current debate. At present, there doesn't exist a recognised functionalist theory of language that puts the tool hypothesis first. Functionalist explanations of language use, language change, and cross-linguistic diversity do not explicitly draw on the tool concept as a central theoretical claim (though see Everett, 2012, 2016, 2017). Instead, "usage-based approaches" figure most prominently in arguments opposing linguistic nativism (Ibbotson & Tomasello, 2016).

This chapter introduces the tool concept as a central claim underlying all functionalist explanations of language use, language acquisition, and its historical development. In particular, the aim is to show that the tool concept can be derived even more directly from the cognitive and behavioral sciences than is hitherto believed. Research in various fields such as developmental and experimental psychology, as well as primatology, is converging on evidence for general learning abilities that lay the foundation of symbolic competences and cumulative cultures even prior to the emergence of language. The indicators of cultural development are here formulated independently of language, but may in a further step be tested in respect to linguistic phenomena. This we do in the subsequent chapters when we turn to the cross-linguistic, typological investigation of reciprocal marking patterns in natural languages.

TOOL USE IN GREAT APES

The first comprehensive insights into the cognitive nature of tool use were gained by the German psychologist Wolfgang Köhler and his work on *Intelligenzprüfungen an Menschenaffen* (1921).[2] Köhler studied a group of some seven captive chimpanzees (*Pan troglodytes*) from 1912 to 1920 at a research station on Tenerife. The rationale behind Köhler's research:

> Given that the anthropoid [ape] shows, under circumstances, intelligent behavior of the kind also found in humans, there is yet no doubt that he will stay far behind man in this respect, i.e. will have difficulties and make mistakes in relatively simple situations; however, this is precisely the reason why his behavior can, under the most simple circumstances, display the true nature of intelligent behavior, whereas at least the adult man, as the object of self-observation, hardly ever performs such simple actions suited for the examination of such abilities for the first time [without prior experience], and has a hard time making sufficiently accurate observations of more complex actions involving himself as subject. (Köhler, 1921, p. 1; translated by the author[3])

Köhler designed experiments that required individuals to overcome obstacles of space. By hindering them to access desirable food objects directly, for example, with their hands or by sheer physical force, he tested his subjects' abilities to develop more creative means to solve nontrivial problems. The intelligence of the behavior was determined by the degree of *insight*, which was measured by observational criteria: According to Köhler, insightful behavior is well distinguishable from "imitations by accident"[4] by exhibiting a continuous and uninterrupted flow of action, forming one integrated whole from beginning to end; random success, by contrast, consists of a mere agglomerate of single movements stringed together in a disconnected sequence of events, which does not make sense as an integrated whole. This basic distinction between insightful reasoning, on the one hand, and trial-and-error learning, on the other, is still adhered to in modern research. "Learning entails combining contiguous experiences," writes the cognitive psychologist Josep Call. "The other way to become a tool user is reasoning," here broadly defined as "combining separate experiences" (Call, 2013, p. 6). Köhler also frequently remarks on the "moment of insight," when the individual suddenly stops short in whatever it has been doing up until that point, usually in vain, abruptly swings around, with apparent purpose, and carries out a new plan of action (Köhler, 1921, pp. 12–13). In modern research, the term *insight* has gone slightly out of use (Mayer, 1995, p. 28), but the described criteria for intelligent decision-making still remain valid for testing the degree of intelligence displayed by a novel behavior (Mayer, 1995; Davidson, 1995; Seifert et al., 1995; Byrne & Russon, 1998; Boesch, 2013; Call, 2013).

It was thus possible for Köhler to expose apes to simple situations in which they were eventually, depending on their individual talent, able to pick out the usefulness of a stick or wooden box in order to reach a desirable goal (usually some food object lying or hanging out of reach). Introducing a subject under suggestive circumstances to the use of sticks or boxes (e.g., by placing them very near the location of the goal) proved only to be the beginning of a series of more complicated test settings. More complicated meant, for example, that the potential tool (stick, box) is placed at a greater distance away from the goal (e.g., on the other side of the room). Under such modified conditions, the situation required that the individual recognize the relevance of the displaced object despite its current spatial distance from the goal; making this connection in turn presupposed some recognition of the functional tool properties of this object in association with the goal, partly based on prior experience. The same was also true of other test settings where subjects had to retrieve knowledge from episodic memory (e.g., one of Köhler's chimpanzees named Sultan having to remember a wooden box lying in the corridor, while being tested in another room; Köhler, 1921, pp. 37–38).

Köhler (1921) repeatedly draws attention to the significance of *the spatial-temporal-visual relationship between tool and goal*: "the best tool easily loses its situational value when it cannot be seen simultaneously or quasi-simultaneously together with the goal region" (Köhler, 1921, pp. 38–39).[5] In the easiest scenario for the individual, it can see both tool and goal "simultaneously," at the same time, so that the functional, associative link between tool and goal becomes apparent within the same perceptual (visual) frame. It gets more difficult for the individual to establish this same link when tool and goal only co-occur "quasi-simultaneously," in the sense that the spatial distance between them does not permit the individual to perceive them both at the same time: looking one way, it sees the goal, looking another (possibly in the opposite direction), it sees the tool. Under such circumstances, recognizing the tool–goal relationship requires the individual to mentally bridge the gap and piece together those two separate impressions. As Köhler's experiments demonstrate, the spatial-visual separation between tool and goal (initially) can pose a serious difficulty for some individuals, even if they know the tool object from previous experience. An example is the behavior of the female chimpanzee Tschego in one of the stick experiments (Köhler, 1921, pp. 22–25): she has already learned to use sticks for fishing bananas from beyond the window bars, with the sticks lying right next to the window. However, in a subsequent trial, she apparently does not perceive the sticks on the other side of the room as tools, but instead keeps ignoring them (even though she does see them). Only when a stick is pushed near her through the bars does she immediately pick it up and reach with it toward the goal.

Being able to use sticks and boxes as simple tools is a necessary prerequisite for being able to perform even more complex forms of tool use; for example, involving the construction of tools or the pursuit of more complex subgoals. Köhler uses the term *tool construction* in a broad sense that also includes basic modifications of an object (e.g., removing hindering material with hands or teeth) (Köhler, 1921, p. 72). A central reason why tool construction is considered more complex than mere tool use is the fact that an individual is required to go more out of its way, invest more time and effort in providing itself with conditions that will enable it to reach its actual goal. In some cases, the individual might spend more time in pursuit of several subgoals (e.g., preparing or constructing a tool) than it will spend time sitting in front of the main goal and reaching for it. Sultan and his peers proved persistent enough in learning, among other things, how to actively prepare tools (by using their hands or teeth), how to construct long double-sticks (by inserting them into one another), how to stack several boxes on top of each other (in order to reach goals hanging several meters above the ground), and how to retrieve tools by using other tools (tools previously only used for reaching the main goal) (Köhler, 1921, pp. 75–77, 90–95, 100–11, 124–31).

In this context, Köhler draws attention to two behavioral patterns occurring as soon as a basic form of tool use has been acquired by the individual: (1) the transfer of tool function from the original object to other objects; and (2) the transfer of tool use to other types of problem settings. Both patterns are part of the same phenomenon, in today's literature commonly described as *means–ends dissociation* (McGrew, 1998, p. 318; Call & Tomasello, 2007; Call, 2013, p. 10; Boesch, 2013), and were observed by Köhler during or soon after the first test trials with a given tool object.

The Transfer of Tool Functions

As soon as the individuals had been given an incentive to use certain tools (stick, box, ladder) in specific situations, they began in subsequent trials with slightly more complicated conditions to pick up all sorts of objects within their reach to replace whatever other object they were missing. The missing tool object (e.g., a stick) was usually still present and within reach for the subject; however, instead of lying close to the goal region it was lying somewhere more distant and out of view for the individual, while focusing on the goal. Under such conditions, individuals would show considerable creativity in selecting replacements. For example, Köhler comments on Koko's obsession at every trial of the stick experiment to pick up whatever object came into his hands:

> Still a few days later he [Koko] uses a big firm piece of cardboard, a rose twig, the rim of an old straw hat, a piece of wire. Anything transportable and longish-shaped turns into a "stick" in this situation, in the purely functional sense of "grasping tool"; one can truly say that the entire moveable part of the field shows a tendency, in Koko's hands, to move in the direction of the goal. (Köhler, 1921, p. 25)

Another example is Sultan's insistent behavior in a ladder experiment: after he fails to put up the ladder underneath his goal, he starts looking for a replacement; to the surprise of his human observers, he picks out *them* as ladder substitutes and attempts (eventually with success) to drag one of the caretakers underneath the goal and climb his back (Köhler, 1921, p. 35). Köhler summarizes: "Over time, all of them were observed to replace box, ladder and table with various other types of objects: stones, spare grids from cage windows, metal drums, wooden blocks, wire rolls; all of them were successfully recruited as chairs or ladders" (Köhler, 1921, p. 35).

In modern primatological research, the same flexibility in the search for tool replacements is observed in wild-living great apes. Such research shows that even under natural conditions, without the interference of human observers, great apes engage in differential tool uses that are informed by abstract knowledge about object affordances and their significance for complex object manipulations. "The material culture of chimpanzees is ubiquitous and diverse," and "a termite fishing probe may be made of bark, stem, twig, vine, or the midrib of a leaf" (McGrew, 1998, pp. 317–18). Moreover, careful observation in the field has shown that such flexible tool choices are not random but subject to foresight and causal reasoning about object–object relations. For example, McGrew (1974) reports of chimpanzees from Gombe, East Africa, that collect and manufacture sticks as fishing poles prior to their arrival at a new feeding site, and observes: "If the tool becomes damaged or coated with soil during use, the chimpanzee may further modify it by biting off worn segments. [. . .] Usually it is shortened until it becomes inefficient, then abandoned and replaced by a new one" (McGrew, 1974, p. 504). Boesch and Boesch (1983), in turn, first reported that the chimpanzees at Taï Forest, West Africa, use either wooden clubs or stones as hammers to crack open different nut species. Careful field experiments further corroborate that individuals are inventive tool replacers if exposed to novel conditions. Thus, chimpanzees at Bossou, West Africa, habitually rely on stones as hammers to crack open oil-palm nuts. In field experiments, however, when the number of available stones is deliberately reduced by the experimenters, especially the young group members become creative in the selection of new functional alternatives: "For example, a young female chimpanzee, who required a hammer stone, started using a wooden club as a hammer [. . .]. Another young chimpanzee used a large wooden branch as an anvil because she had no access to anvil stones" (Matsuzawa, 2011, p. 160).

The Transfer to Novel Problems

The same degree of flexibility can also be observed in the other direction; that is, with regard to the same tool object (means) being used to reach different ends. Thus, Köhler (1921) already observed that once a given tool use had been introduced to an individual in a specific context and once its use had been consolidated through simple variations of the original test setting (e.g., by placing the tool at greater distances from the goal region), individuals invariably began to transfer this tool use to other situation types. "As might be expected, variations of the goal or its location generally remain without any significance once the use of a stick has been learnt" (Köhler, 1921, pp. 25–26). This transfer applied to novel test situations as well as everyday activities outside of test trials, and it occurred quite naturally, that is, without any explicit encouragement by the human caretakers.

> The interaction with things, as can be daily observed in the [captive] chimpanzee, can almost exclusively be categorized as "play." Once a given tool use or something of the kind has been acquired out of the "necessity" of a test setting, one can be quite certain to find the new element of behavior resurface in play, where it does not possess any immediate "advantage," but is a mere expression of joy. Likewise, it can easily happen that elements of behavior first acquired in random play, which the chimpanzee performs in interaction with objects, will on occasion turn out to be of great practical advantage [. . .]. (Köhler, 1921, p. 49)

With respect to the use of sticks, Köhler further observes: "The stick is a kind of universal tool for the chimpanzee; in almost any situation one can make do with it. Once its use had been adopted and had become common knowledge [within the group], its function became ever more diverse each month" (Köhler, 1921, p. 52). The following is a list of invented stick uses established among these apes as ways of passing the time (see Köhler, 1921, pp. 52–61):

- using sticks, pieces of wire, straws or poles for pulling things through the fence;
- using blades of grass or twigs to reach through a wooden lid and to spoon into a sewage canal; using a stick to leverage the wooden lid out of its foothold;

- using straws, twigs, and so on during meals in a playful manner; for example, in order to bring water to the mouth (after having already quenched one's thirst by mouth), or putting straw into the mouth and chewing it together with fruits (when one is not particularly hungry);
- using straws or little twigs for fishing ants (instead of just using mouth and tongue);
- using sticks to dig in the ground, either with or without distinct purpose (e.g., finding roots);
- using sticks to satisfy one's curiosity toward inanimate things like fire or animals like mice and lizards; being both fascinated and scared at once; using stick as weapon when animal unexpectedly jumps toward ape;
- using stick as gripping tool when reaching for slightly electrical food basket (experiment); using the same stick as weapon after receiving electrical shock from basket (hitting basket with both hands on the stick);
- using stick in jest fight; however, abandoning stick as soon as play is turning into real fight;
- using sticks or hard wire to play with (tease) chickens.

Two observations are particularly striking when scanning through this list of social group occupations:

First, the use of stick-like objects became transferred to all manner of situations in which both the tool function and individual goal-orientation had not much in common with the original test setting. Activities such as reaching with sticks through the fence (for plants), a wooden lid (for sewage), or to the other side of a crossbar (for ants) may still be of a similar type (i.e., reaching for out-of-reach objects); however, activities like using sticks for digging in the ground, leveraging a wooden lid, drinking water, fishing for ants within reach, and teasing chickens are quite different from one another in both technique and purpose. The different situations to which the use of stick-like objects becomes transferred bear no simple resemblance to one another, neither in terms of tool function, nor purpose, nor in terms of material tool object. In some cases, sticks are not even necessary in order to reach some goal (e.g., digging earth, dipping water, inspecting animals), nor are the activities really aiming for a distinct goal. Instead, Köhler observes that social group dynamics seemed to play an important role in promoting the temporary occurrence of such tool uses. In the following passage, he writes about the occasional practice of "ant dipping" among all group members:

> In this case, too, it seems that the sportsmanship was greater than the appetite for ants, since there were many locations where the ants could have simply been licked off, and under otherwise unchanged conditions, the most beautiful column of ants would simply be ignored if the fashion, at the time, tended another way. However, if the fashion did tend this way, one could observe all animals

of the station sitting one next to the other along a column of ants, each with a piece of straw [in his hand], like a row of anglers along a current. (Köhler, 1921, pp. 55–56)

Second, the choice and manner of tool use emerged quite naturally from the activities at hand; that is, the circumstances in each case dictated *which objects* could or must be used in order to successfully perform certain tasks, and the same circumstances also dictated *in which manner* that tool use had to be performed. For example, reaching through a fence to target plants requires a long firm stick, yet one that still fits through the fence. Reaching through the cracks of a wooden lid to target some liquid requires the use of little twigs or grass stems. And the leveraging of the lid in turn demands a strong, pole-like object. Also, removing a wooden lid from a sewage canal requires other movements than removing earth from the ground, or fishing for ants. Again, all these differences in object choice and manner of usage clearly emerge out of the given context, together with the goal-orientation. As a result, not much reminds one of the original test settings in which this tool use had first been acquired.

In modern primatology, this capability of flexible transfer of means still remains an important criterion of intelligent tool use. "The same raw material serves multiple functions," notes McGrew (1998) with regard to wild-living chimpanzees: "A leaf may be a drinking vessel, napkin, fishing probe, grooming stimulator, courtship signaler, or medication" (p. 318). Boesch (2013) also highlights the variety of tool repertoires in nonhuman primates. "Flexible tool use, the use of different types of tools in different contexts and to reach different types of rewards, has been observed more frequently in chimpanzees and humans than in any other animal species" (Boesch, 2013, p. 25). Moreover, field experiments on chimpanzees in the Bossou region, West Africa, demonstrate the challenge posed by the initial transfer of dedicated means to novel ends (Biro, 2011). In such studies, group members are exposed to large quantities of unfamiliar nuts, while only few or none of the familiar nut species are made available to them. In such cases, only some, often juvenile group members are observed to go creative and to apply their learnt technique of using stones to crack open the unfamiliar food items. Through observation, other individuals then gradually follow their lead and adopt the same pattern of transfer (Biro, 2011, pp. 170–71). Thus, the creativity to generalize acquired means to novel ends is a conspicuous and exceptional feature in the behavior of great apes.

TOOL INTELLIGENCE AND SOCIAL TRANSMISSION

Based on his empirical observations on problem-solving in captive chimpanzees, Köhler (1921) developed a theory of intelligent behavior that still forms the basis of contemporary theories of complex behavior of both humans and animals. Köhler's experiments with captive chimpanzees demonstrated how even the most basic forms of tool use have to be learnt by—what he called—*insight* under specific problem settings. His qualitative observations proved that naïve individuals do not simply jump at a solution by being motivated and able to visually perceive potential tool objects. Instead, it often took them several minutes or even trials before functionally adequate associations were made between some tool object and a given target area. Moreover, individuals showed great variation with respect to their ability to recognize and select potential tool objects: some individuals were extremely quick; others needed several trials before their first insightful behavior or sometimes even failed entirely to become confident tool users in a certain domain. Nevertheless, once such basic associations had been established, they stayed with the individuals and became part of their habitual behavioral repertoire.

Current theorizing about intelligent tool use is still based on Köhler's early gestalt-psychological assumptions (Mayer, 1995). These basic assumptions are that individuals have *intentional* goal orientations and are able to *voluntarily* plan and control their behavior in a context-dependent manner (Call, 2013; Boesch, 2013). Moreover, this ability to interact with the environment voluntarily is characterized by a propensity to *manipulate* objects (Call, 2013, p. 15) and to apply this ability to an extended *awareness* of object–object relations (Campbell, 2011; Call, 2013). In other words, nonhuman primates solve problems by active, playful engagement with their environment and by inventing means (individually) to bring about physical changes of states within this environment. The ability to reason, that is, the ability to quickly retrieve, compare, and combine different memories and to further translate them into complex actions, determines the degree of insight and the complexity of behavior displayed by an individual (Byrne, 1995; Call, 2013). An increasing amount of actions planned in this way are not directed toward the main goal but toward subgoals that merely serve as intermediary steps for reaching the main goal. At least to a minimal degree, then, such problem solving is already *hierarchically organized* (Gibson, 1993; Byrne & Russon, 1998). In modern literature, another term used to characterize this type of action planning is *recursion*.[6] However, a distinction is drawn between recursion as a theoretical or mathematical construct and recursion as a subject's plan of action: in theory, as a formal construct, recursion can lead to endlessly complex action sequences; in practice, as a mental construct, however, such endlessly complex behavior is not possible:

> In a real system, such as the mind of a gorilla or a human, short-term memory capacity might set a limit on how deep a hierarchy it could expand without getting into a muddle. [. . .] Integrated groups of these basic [behavioral] elements,

> assembled together during the animal's interactions with the world, form higher-level units of behaviour, and this process of hierarchical grouping can continue to arbitrary complexity, though in practice it may be limited by the available mental capacity. (Byrne & Russon, 1998, pp. 673, 675)

There are indeed strong constraints on short term and thus working memory in great apes, which limit their ability to construct hierarchically organized action sequences (Köhler, 1921; Gibson, 1993; Byrne & Russon, 1998; Read, 2008). Scholars generally agree that by comparison "humans possess a greater degree of brain-size mediated information processing capacity and apply this capacity to their tool use, language and social behavior" (Gibson, 1993, p. 251; see also Finlay et al., 2001; Marino, 2006). Some even go further and argue "that the chimpanzees (*Pan troglodytes* and *Pan paniscus*) have insufficient working memory capacity to enable recursion to be part of their cognitive repertoire" (Read, 2008, p. 678). Yet, even in humans, working memory is limited so that this *quantitative* difference in mental capacities and hierarchical organization of behavior does not necessarily warrant the assumption of a *qualitative* difference in underlying cognition (Gibson, 1993; Bryne & Russon, 1998; Savage-Rumbaugh et al., 1998).

Another important characteristic of tool-using behavior, which brings us closer to an analogy with human language, is its *social transmission* within the group. Even Köhler (1921) already observed that some of the behavioral variants involving sticks or other materials occasionally spread among all group members. He characterized these behaviors as "behavioral fashions" because of their opportunistic and often only periodically maintained character. They came and went in tidal waves, depending on individual whims and the social dynamics within the group. In other words, their spreading and maintenance seemed more related to social factors than external factors to do with subsistence or technical necessity in interaction with the environment.

Careful experimentation since then has complemented this picture in important ways. Comparative psychologists and primatologists have carried out experiments studying both the propensity and ability in great apes (and other primates) to socially transmit simple and complex behaviors within the group (see reviews in Whiten et al., 2004, 2009). What has been tested in such studies is twofold. On the one hand, the question was to what extend nonhuman primates rely on social transmission in the first place to acquire and maintain behavioral repertoires in relation with their environment. The evidence shows that chimpanzees, like other primates, are extremely social and tend to copy arbitrary behaviors in positive correlation with other social factors relating to social proximity, age, or dominance (Whiten et al., 2005; Bonnie et al., 2007; Horner et al., 2010). Juveniles tend to copy their elders; individuals with close social ties tend to copy each other more often than individuals not living in close, friendly proximity. Moreover, dominant relations dictate who copies from whom, and whether useful variants of behavior are adopted within the group or not. In such cases, social transmission remains conspicuously independent of nonsocial, functional evaluations of invented behaviors. Hence, ecological factors do not play a decisive role here, which to some extent of course owes to the artificial test setting of such experiments.

On the other hand, such experimental studies have tackled the question of how *skilled* nonhuman primates are in copying complex behaviors observed in others. Such testing relates to the still ongoing, contentious debate on *true imitation*. The question relates to whether nonhuman primates can understand the intentions and goals of others based on which they can acquire complex object manipulations requiring nontrivial tool intelligence. Some scholars tend to argue against this hypothesis and assume that nonhuman primates simply *emulate* solutions to complex problems by figuring out problems individually based on the effects their manipulative activities have on their physical surroundings (Tomasello et al., 1987; Tomasello, 1990; Tennie et al., 2009). Others, in turn, oppose the emulation hypothesis (Whiten et al., 2004, 2009; Boesch, 2012). They point out the unlikelihood of individuals inventing similar behaviors independently of one another without attending to what others do and understanding why they do it. Thus, along these lines, studies have shown that great apes are faithful adopters of special food-processing or food-retrieving techniques introduced by other group members (Whiten et al., 2005, 2007; Hopper et al., 2007; Vale et al., 2017). The latter are taught independently by human caretakers to use a special technique; afterward they are released as *models* into the group and provided with continued access to the novel food resource requiring the special technique. Under such conditions, the surrounding group members end up with the same overall technique introduced by their model conspecific (thereby leading to the introduction of alternative techniques in separately tested groups). Such evidence suggests that social transmission is not only limited to emulative learning but also involves a basic understanding of the tool-using intentions of others: "the picture emerging is that apes may develop a portfolio of social-learning processes that are differentially activated in different contexts and in relation to different functional requirements" (Whiten et al., 2004, p. 46). These findings are important as they highlight the possibility to transmit behavioral innovations within the group and to build on these over subsequent generations as cumulative cultural traditions.

PATTERNS OF CULTURAL VARIATION IN TOOL USE

Field observations on great apes have corroborated the social transmission hypothesis of tool use independently of the evidence presented in experimental studies. In particular, long-term studies on wild chimpanzees (*Pan troglodytes*) living in different parts of Africa have revealed the existence of very different behavioral repertoires practiced in different communities. Since in many cases, the specific form and function of these behavioral patterns cannot be explained either in terms of genetic affiliation or environmental conditions, primatologists have come to systematically classify these patterns as *cultural variants* of behavior (McGrew, 1998; Whiten et al., 1999, 2001; see also Whiten, 2005, 2017, who speaks of a "second inheritance system"). As a result of group-specific analyses as well as cross-group comparisons, it is thus possible to provide very detailed evidence of the existence of complex and diverse tool cultures among great apes.[7] The best proof of the cultural origin of group-specific behaviors is given on the basis of *geographic variation* that remains independent of either genetic or environmental explanations. In the following, we illustrate this comparative method with a focus on nut-cracking behavior of wild chimpanzees. Equipped with the results, we then draw *the analogy to language use* in chapter 4 and following.

In principle, nut-cracking behavior, that is, the use of dedicated tools like stones or pieces of fallen wood to crack open nuts, may be encountered in different chimpanzee communities living in different parts of Africa. A central question relates to the precise origin of such behavior: is it innate, or did its form and function arise from group-specific innovations and subsequent historical processes carried by social transmission within and across the generations? The predictions are as follows: If the behavior is innate, chimpanzees in general should possess the basic disposition to use stones (or other stone-like objects) in order to crack open nuts or other available types of hard-shelled fruit. That is, all chimpanzees should possess the basic knowledge of how to use tools in order to crack open nuts independently of the social environment in which they grow up. The tool-using competence would thus not be based on socially transmitted knowledge within local communities, but on genetically predetermined associations within the individual.[8] As a result, it should not matter which chimpanzee community we choose to study: given a natural environment that supplies local groups with a sufficient amount of nut-bearing plants and a sufficient number of potential tool objects (stones, wood), dedicated nut-cracking behavior of a certain form and function—more or less identical across groups—should occur. If, on the other hand, the behavior is cultural, we should expect to find patterns of geographic variation between groups. That is, chimpanzees living in different communities should practice different foraging skills and these either do or do not include the cracking of hard-shelled nuts with dedicated tools (McGrew, 1998, p. 305). In this view, the spectrum of variability across groups could be immense: it could range from groups with dedicated nut-cracking techniques applied to many different nut species, to groups with techniques applied to only very few nut species, to groups with less dedicated foraging skills like cracking nuts with bare hands, to groups with no nut-cracking behavior at all. Of course, in cases where dedicated nut-cracking skills are absent, it should be possible to exclude either genetic or environmental factors as alternative explanations.

Primatologists from the field hold that the latter assumption alone is true, that is, that we are dealing with cultural variants of behavior. They base their evidence on both group-specific observations and cross-group comparisons. To begin with, in a large-scale comparison between groups living in different parts of Africa, it is noted that nut-cracking behavior involving stone/wood "hammer" and "anvil" is only practiced by some communities living in West Africa (e.g., Bossou, Taï) (Sugiyama & Koman, 1979; Boesch & Boesch, 1983), while it is found to be absent in communities living in Central or East Africa (e.g., Mahale, Gombe, Lopé) (McGrew et al., 1997; Ghiglieri, 1984; Nishida & Uehara, 1983; Goodall, 1986).

Moreover, it is noted that some communities that lack dedicated nut-cracking skills nevertheless do employ other "percussive technologies" in application to some hard-shelled fruits. Thus, for example, chimpanzees from Gombe do not use dedicated tools (stones, wood) to crack open nuts, but they do employ a smashing technique on hard surfaces for opening a special type of hard-shelled fruit (*Strychnos*). In other groups like Mahale or Lopé, it seems, even such smashing techniques are absent. Importantly, in such cases, the absence of nut-cracking behavior cannot be explained in terms of environmental conditions. Hence, as noted by Goodall (1986), chimpanzees from Gombe do have oil-palm trees (*Elaeis guineensis*) available to them but do not try to crack their nuts open.[9] Likewise, as shown by McGrew et al. (1997), chimpanzees at Lopé live in a forest very similar to the one inhabited by chimpanzees from Taï Forest, including several identical nut species like *Elaeis guineensis*, *Coula edulis* and *Panda oleosa*; but, yet again, they are not observed to crack open nuts.

In a next step, on a smaller-scale level, researchers have investigated the distribution of nut-cracking behavior *within* different regions of West Africa. Over the years, two important observations have been established. First, the practice of dedicated nut-cracking is only limited to chimpanzee communities living in the western parts of West Africa (Sierra Leone, Liberia, Guinea, Western Côte D'Ivoire), basically just the regions west of the Nzo-Sassandra River divide in Côte D'Ivoire. This finding contradicts the hypothesis that the occurrence or nonoccurrence of dedicated nut-cracking behavior is correlated with genetic factors such as belonging to the same subspecies.[10] On the other hand, as formulated in Boesch et al. (1994), it corroborates predictions made by the cultural diffusion hypothesis, according to which the nut-cracking skill was invented at a time when the chimpanzee communities living on either side of the Nzo-Sassandra region had become geographically isolated from each other into western and eastern living communities. This would explain why the skill did not spread east by social transmission beyond the Nzo-Sassandra River divide, which today continues to be impassable for chimpanzees living on either side.[11]

The second observation concerns the precise form and function of the nut-cracking skill practiced in different communities west of the Sassandra River. Thus, a comparison between the nut-cracking techniques practiced at two long-term study sites—Taï Forest, Côte D'Ivoire, and Bossou, Guinea—reveals that chimpanzees from each community stick to both a *different technique* and a *different application pattern*. Chimpanzees from Taï Forest rely on stones or pieces of fallen wood as "hammers" and on embedded stones or broad-shaped tree roots as "anvils." Moreover, they apply this technique to at least five different nut species (Boesch & Boesch, 1983). Chimpanzees from Bossou, on the other hand, are known to use movable stones as both "hammer" and "anvil," thus not (or only sometimes) relying on embedded stones or tree roots for the latter function. Moreover, they are never observed to use this special technique anywhere else than in application to oil-palm nuts (*Elaeis guineensis*), the one nut species never touched by chimpanzees in Taï Forest (ca. 250 km away) (Sugiyama & Koman, 1979). These findings are compatible with various predictions made by the cultural invention and diffusion hypothesis: one possibility is that each type of dedicated tool use was invented and socially transmitted independently of each other; another one is that both nut-cracking techniques share a common history prior to the diversification into separate communities. In the latter view, present-day differences are only due to more recent historical developments, again based on local innovation and social transmission.

We can summarize and depict some of the previous findings schematically as shown in Figure 3.1. As can be seen in the figure, chimpanzees from Taï Forest even engage in more than just one type of percussive technique; while in application to most nut species, they employ the just mentioned hammer-and-anvil technique, they engage in mere manual smashing on hard surfaces with hard-shelled fruits like *Strychnos* and *Treculia*. This latter technique is also used by chimpanzees from Gombe, East Africa, that otherwise lack a hammer-and-anvil technique.

Finally, large-and-smaller-scale cross-regional comparisons between groups (as outlined above) are further complemented by small-scale comparisons between adjacent, or almost adjacent, chimpanzee communities within West Africa. In the Bossou region of Southeast Guinea, such comparative surveys in neighboring territories have over the past decade been specifically targeted at the systematic documentation of nut-cracking behavior (Matsuzawa et al., 2011). As is discussed and summarized in the relevant literature (Humle & Matsuzawa, 2004; Biro, 2011; Humle, 2011b; Carvalho, 2011),

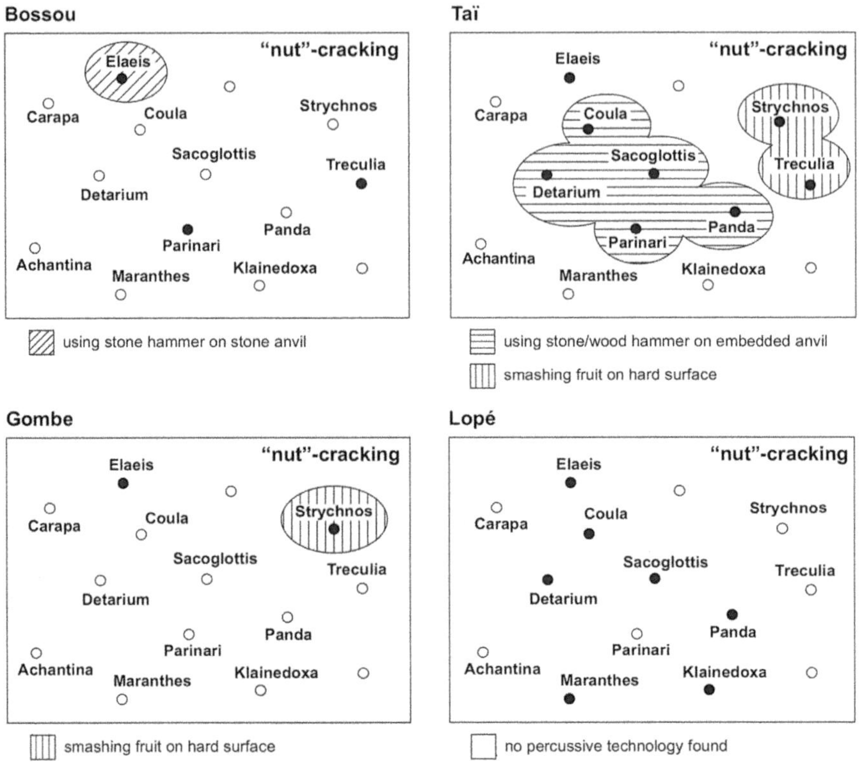

Figure 3.1 Geographic Distribution of Nut-Cracking Behavior Among Wild Chimpanzees (Part I). The placement of dots in each "nut"-cracking frame is identical, with each dot corresponding to a specific type of hard-shelled nut/fruit species (except the dot labeled Achantina, which denotes a shelled snail species; more about this in Figure 3.2). In each frame, the prominent black dots denote nut species that are known to be available within the home range of the respective ape community; the less prominent empty dots, on the other hand, denote nut species that are either not available or at least not known to be available to the chimpanzee community at hand. The shaded entries around black dots in each frame indicate the type of technique(s) employed by chimpanzees from the community in application to one or several different nut/fruit species within their home range. Different shadings indicate different techniques; same shadings indicate similar techniques. Black dots that are not circled by a shading thus indicate nut species left untouched by the chimpanzees of a given community. All references from which this information was drawn are listed in Table 12.1, Appendix C. *Source*: Author created.

four chimpanzee communities living in four adjacent territories (Bossou, Seringbara, Yealé, Diécké) do not practice the same foraging skills on nuts. As was outlined earlier, chimpanzees at Bossou only feed on oil-palm nuts (*Elaeis*) by means of movable stone hammers and anvils. Other available nuts like *Parinari*, on the other hand, are left untouched. The chimpanzees living directly adjacent to the Bossou region near the village of Seringbara, on the other hand, do not employ any dedicated nut-cracking at all, though several nut-bearing trees (*Elaeis, Detarium, Parinari*) and potential tool objects (stones, rocky outcrops) are freely available to them. Nevertheless, like chimpanzees at Gombe (East Africa), chimpanzees at Seringbara are not entirely without percussive technology. In this context, Koops et al. (2010) report indirect evidence of a dedicated smashing technique involving either stones or sticks as "cleavers," or stone outcrops as "anvils" for the fracturing of *Treculia* fruits. (Recall that the same type of fruit is also smashed by Taï chimpanzees [see Figure 3.1], however *not* with stones or sticks as cleavers.) The other adjacent community, living a bit further away on the other side of the Nimba Mountain range near Yealé, is again more similar in behavior to the one at Bossou—though details about the exact nut-cracking techniques in this area are yet unknown (Humle, 2011b, p. 270). Here, chimpanzees also use movable stone hammers and anvils for cracking open oil-palm nuts (*Elaeis*). However, in addition, they also employ the stone hammer and embedded anvil technique (as in Taï), indirect evidence suggesting that they extend this behavior across a wide range of different fruit types such as *Coula*, *Carapa*, and *Strychnos*. (Recall from above that the latter were also accessed through the smashing technique by the chimpanzees at Taï, Côte D'Ivoire, and Gombe, East Africa.) Further, indirect evidence in the Yealé region suggests that chimpanzees even apply this technique to feed on shelled *Achantina* snails (Matsuzawa & Yamakoshi, 1996). Finally, data from the Diécké Forest, located some 100 km southwest of Bossou, further complements the picture of small-scale cross-regional variation (Carvalho, 2011, pp. 305–10; Biro, 2011, pp. 166–68). Here, too, dedicated nut-cracking involving stone hammer and embedded anvil exists. However, its application pattern only partly overlaps with that at Yealé. Chimpanzees pound Coula and Panda nuts, but apparently do not extend this technique to oil-palm nuts (*Elaeis*) and *Strychnos* fruits. Consequently, there appears to be no overlap at all, in either technique or application pattern, with the practice at Bossou. Figure 3.2 summarizes these findings.

As these small-scale cross-regional comparisons illustrate, even chimpanzee communities living directly adjacent to one another do not necessarily share the same patterns of behavior (dedicated tool use) in application to similar contents (here, nut/fruit species). This is true even if in other respects neighboring communities still resemble one another in local customs (e.g., other types of foraging, nest building)—suggesting that due to their geographic vicinity they are yet culturally related to one another.[12] Of particular interest is the observation that a complex skill like dedicated nut-cracking as it is found at Bossou can be totally absent in one neighboring community (Seringbara) but still be present in another one living just a bit further away (Yealé). This finding is again only compatible with predictions made by the

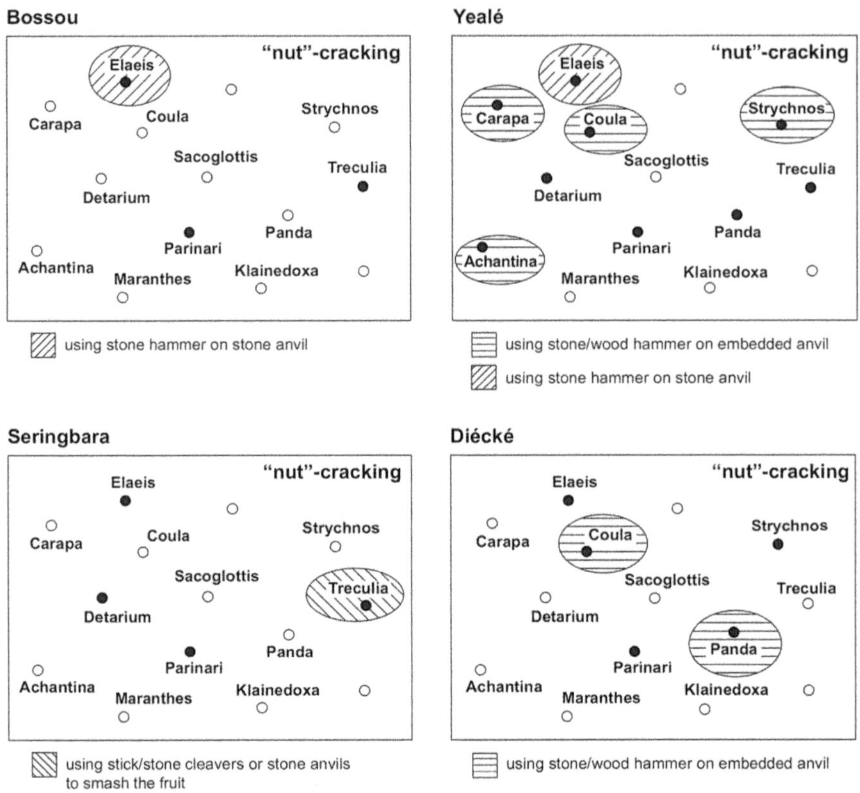

Figure 3.2 Geographic Distribution of Nut-Cracking Behavior Among Wild Chimpanzees (Part II). As in Figure 3.1, the "nut"-cracking frame for each community offers the same "possible space" of application, indicated by single dots and the genus names of fruit-bearing plants (excepting the dot labeled Achantina, which denotes a type of shelled snail, apparently cracked by chimpanzees living at Yealé). Black dots in each frame indicate the range of nut-bearing plants available within a given territory; empty dots, on the other hand, indicate their (potential) absence. The shaded circles around black dots indicate the used cracking or smashing technique practiced within the community, labeled underneath each frame. (The Bossou diagram is a repetition from Figure 3.1.) All references from which this information was drawn are listed in Table 12.1, Appendix C. *Source*: Author created.

cultural hypothesis: for example, based on local patterns of migration and social interchange between different groups, it is possible to explain why skills may spread to some communities but not to others; likewise, the theory can predict that skills may go extinct as populations migrate from one territory into another; or, furthermore, that local communities might not possess a certain skill because they originally migrated to the region from an area further away, where the skill had not been invented; and so on. Of course, with respect to the pattern of variation currently encountered in the Bossou region, only some of these predictions made by the cultural hypothesis will be appropriate; however, as the precise histories of the local communities are at present unknown, the explanations must remain speculative.

PREDICTIONS FOR CULTURAL VARIATION IN LANGUAGE

In summary, we can make several general observations with regard to *patterns of variation* in tool-using behavior among wild chimpanzees. From each generalization, we also derive a general prediction with regard to the study of language and, in particular, the typological investigation of reciprocal markings across natural languages. These grammatical devices will serve as a case study just as nut-cracking behavior is here taken as an instance of tool use among great apes. The predictions formulated next will be tested on natural languages in chapters 4 and following.

Different Techniques

Chimpanzees living in different communities acquire and practice different technical skills for feeding on important food resources. In some cases, dedicated skills involving complex action sequences and specific tool objects (e.g., stones, sticks) are the only means to access a given food source, or else they at least facilitate the harvesting (Yamakoshi, 2001). As the cross-populational evidence shows, chimpanzees living in different social groups by no means always possess the same kind of dedicated skill in order to access such hard-to-get food sources. Instead, their technical repertoires vary across groups.

Furthermore, even chimpanzee groups engaging in dedicated techniques differ from each other with respect to the precise practice of this skill: chimpanzees at Bossou use both movable stone hammers and anvils to crack open nuts, while chimpanzees at Taï Forest rely on movable stone/wood hammers in combination with embedded stones or tree roots as anvils. That is, Taï chimpanzees are never observed using movable stone anvils (though Bossou chimpanzees occasionally do employ embedded anvils instead of movable ones). Even between communities where essentially similar forms of nut-cracking behavior are practiced, differences in behavior exist. For example, neighboring communities in Taï Forest show different preferences for stones and wooden clubs as functional tools for nut-cracking (Luncz et al., 2012). Moreover, chimpanzees from Diécké seem to develop strong preferences for individual stone tools and to transport these from one nut-cracking site to the next (Carvalho, 2011, pp. 308, 310). This transport custom differs from other sites such as Bossou or Taï. While chimpanzees from Taï sometimes transport stone or wooden tools from several hundred meters away to a nut-cracking site, they do not transport tools from one nut-cracking site to another (Boesch & Boesch, 1983).

Finally, chimpanzees from communities that do not engage in dedicated nut-cracking (i.e., involving the combined use of stone/wood "hammer" and "anvil") may still possess other types of dedicated technique in order to crack open hard-shelled fruits. Examples are the fruit-smashing on hard surfaces among chimpanzees at Gombe, East Africa, and the fruit-fracturing with wooden "cleavers" among chimpanzees at Seringbara, West Africa. Due to the fact that these other techniques tend to involve (1) other tool sets, (2) different motor patterns, and (3) different types of hard-shelled fruits, they are usually classified more generally as variants of "percussive technology" rather than dedicated "nut-cracking" behavior. Nevertheless, as the occasional overlaps in application pattern suggest (see Figures 3.1 and 3.2), they may yet serve similar functions and thus fall within the same functional domain.

It follows that the specific form and function of nut-cracking behavior in local communities must be the cumulative result of cultural-historical developments, which in turn are based on single innovations and subsequent social transmission within the group.

Prediction 1

Reciprocal marking in natural languages differs in cross-linguistic perspective. Natural languages employ different structural, morphosyntactic means to encode reciprocal meanings (situation types). Such variation is unpredictable across languages in the sense that it is not possible to predict what type of syntactic and/or morphological marking will be habitually employed in any chosen speech community. Moreover, even where languages employ similar means, that is, construction types, the forms still differ in detail and do not conform to a single, cross-linguistically valid rule of grammaticality. It is further predicted that languages are likely to employ a broad spectrum of constructional means in reference to reciprocity. Some of these constructions cover the core domain of what is generally defined as reciprocal meaning, while others only overlap with this functional domain at its periphery.

Different Application Patterns

The tool-using behaviors of wild chimpanzees do not just vary with respect to the types of dedicated technique but also with respect to the specific application patterns associated with each type of dedicated skill. In some cases, this variation can be explained by environmental factors (e.g., absence of nut/fruit species within the home range), in others it is explained by functional factors (e.g., inadequacy of available technique); however, in others, in can again only be explained in terms of cultural-historical development.

Best proof of the cultural hypothesis is given in such cases where chimpanzees living in different communities have the same kind of food and potential tool resources available to them, yet do not engage in similar behaviors in order to exploit them. Hence, for example, while Bossou chimpanzees only employ hammer and anvil in order to crack open oil-palm nuts, thereby ignoring other available nut species like *Parinari*, Taï chimpanzees ignore oil-palm nuts, while employing the hammer-and-anvil technique to feed on five other nut species, including *Parinari*. Similarly idiosyncratic are the application patterns of the hammer-and-anvil technique among the chimpanzees from Yealé and Diécké. The former apply this technique even in contexts where other nut-cracking communities, like Taï, rely on a different technique (e.g., the smashing of *Strychnos* fruits); furthermore, they even seem to go so far as to apply it to food sources other than nuts/fruits such as shelled snails of the genus *Achantina* (Matsuzawa & Yamakoshi, 1996). Among chimpanzees from Diécké, in turn, the application pattern resembles that of Taï Forest by including *Coula* and *Panda* nuts, but also differs from it by excluding nuts from *Parinari*.

Such examples demonstrate that even the same types of dedicated technique practiced in different communities are bound up in different distributional patterns. That is, even though under purely functional (= technical) considerations we would allow a given technique, such as the hammer-and-anvil technique, to occur in a wide range of different contexts, we are still faced with the empirical fact that within single ape communities this logical space of conceivable applications is never fully realized. Instead, the application pattern in each community is only patchy, group-specific, and clearly limited to only some of the available food sources. This observation can only be reconciled with the assumption that the invention as well as subsequent extension process of a given skill is context-dependent and subject to group-specific innovation and social transmission events. To conclude, no explanation other than in terms of cultural-historical development is conceivable with respect to the cross-group variation in application pattern exhibited by functionally equivalent techniques.

Prediction 2

The logical space for reciprocal marking is greater than the empirically occupied space of reciprocal marking in individual languages. The conventionalized distribution and productivity of reciprocal markings is thus limited and language-specific. Some reciprocal markings are highly productive, extending across a wide range of reciprocal situation types. Others are less productive and are limited by constructional (syntax-related) and/or semantic and pragmatic (meaning-related) factors. Some reciprocal markings may even be entirely unproductive and limited to a very specific set of lexically defined constructions.

The Lack of Dedicated Techniques

The rationale of intelligent tool use is immediately accessible to a human observer: we *do not* expect tool use to occur in contexts where it is not necessary, that is, where obstacles can be overcome by more direct means than complicated tool use; conversely, we *do* expect tool use to occur in contexts where dedicated object manipulation is necessary in order to overcome an obstacle. In other words, tool use is not so much aimed at the goal or target itself (e.g., the kernel inside a nut), but rather at a specific problem setting connected with that goal through circumstances. In short, tool use is *problem-oriented*. In this context, Boesch and Boesch (1990) provide useful observations with regard to the seasonal flexibility with which chimpanzees at Taï Forest crack open *Detarium* nuts with or without the help of dedicated tool use: "*Detarium* nuts are eaten at different periods, either using a tool or not. [. . .] tool use seems to be more efficient, although both the nuts and the hammer have to be transported to an anvil. Some individuals may use both techniques, but at the end of the *Detarium* season most of them eat nuts without the help of a tool" (Boesch & Boesch, 1990, p. 91). This implies that dedicated tool use can be absent in many contexts where the same goal is pursued, but under different, less problematic circumstances. Chimpanzees from a community like Lopé, Central Africa, that do not possess a dedicated nut-cracking skill (McGrew et al., 1997) (see Figure 3.1), could thus still access nuts (e.g., *Detarium*) during those periods of the nut season where the obstruction of the nuts is possible even without a dedicated technique.

This last aspect raises interesting questions with regard to the importance of dedicated tool use and, by implication, the lack of it. The lack of dedicated tool use can, under some circumstances, truly limit the possibilities for individuals to reach certain goals. If chimpanzees at Taï Forest had not developed the skill to crack open nuts with stone hammer and anvil, it would not be possible for them to access nuts such as *Detarium* during the earlier stages of ripeness when the nuts are still too hard unless opened with a tool. Under certain ecological conditions, the lack of such knowledge could thus limit their possible scope of action and thereby decrease their overall chances of survival.

On the other hand, the *functional necessity* and value of dedicated tool use does *not* necessarily show to an equal degree in all the different contexts in which it is practiced within the group. That is, it may also resurface in contexts where other groups reach the same goals without a dedicated technique. In this context, it may be useful to distinguish between the following two aspects: (1) the functional motivation and value to *invent* a given tool use, which consists in the technical necessity to overcome certain obstacles under specific conditions of action; (2) the functional motivation and value to *maintain* a given tool use, which is the cumulative result of social transmission within and across the generations. As behaviors spread among group members, social group customs emerge, which include both functionally necessary and seemingly unnecessary, that is, more optional applications of dedicated tool use. However, even in the case of more optional applications of tool use, the generalization of tool practices within the group may be more economical than the sparse application of tool uses to selected instances. It may thus be more economical to generalize innovations across a wide range of everyday contexts than to restrict such knowledge to only a few, technically challenging contexts. In other words, the cost of acquisition and maintenance of a novel behavior may only be properly outweighed through its subsequent transfer and generalization, rather than by its restriction to only a few specific contexts of application.

Prediction 3

Not all languages have dedicated reciprocal constructions. Instead, some rely on more indirect means, and suggestive context settings in communication, to express the reciprocal relation. It is further predicted that languages differ with respect to the degree to which their reciprocal markings are generalized. Some languages make reference to certain reciprocal meanings without dedicated means; others make the same references relying on such more dedicated means. Even in such cases, however, usage-based factors can be invoked to explain such broad extensions of dedicated (structurally more costly) means.

The Functional Properties of Tools

Different tool objects have different potentials for transfer. For example, different types of sticks, stones, or other materials (bark, leaves, etc.) have quite different physical affordances that form the basis of their functional properties as tools. Sticks are more or less rigid, longish, pointed and thereby afford manipulations like poking, hitting, reaching, and leveraging. Stones, on the other hand, are rigid, heavy, roundish and thereby permit manipulations involving percussive strategies. Hence, stones are highly efficient when cracking nuts; however, less so when the intention is to reach for the insides of a beehive. Similarly, using a piece of bark for dipping termites or ants from a surface might be useful, but that same tool object does not function efficiently for fishing insects out of a termite mound. In other words, the choice of tool object is never just determined by the simple intention to forage on food, but by the type of problem, the type of obstacle to be overcome under given circumstances.

This basic property of objects as tools and the concomitant logic of problem-oriented transfer allows one to make an important generalization. Namely, functional categories characterized by natural tool uses are only indirectly correlated with primary perceptual categorizations of the environment. Tools are not directed at specific goal areas, targets, and themes, such as specific food resources like ants, termites, bees, nuts, or fruits. Nor are they specifically directed at certain physical or social desires like hunger, curiosity, anger, or fear. With regard to either, their categorizations are only crude, vague, and nonspecific. Rather, tools are directed at specific problem settings. They *refer* to abstract, physical-causal object–object relations that are susceptible to manipulation by external means. Under such higher-order structural perceptions, problem settings from quite different goal areas or themes become subsumed under the same labels (concepts) of tool use. Hence, as we saw in the nut-cracking examples, quite different nut species are subsumed under the same dedicated percussive technique, because they share the hard-shelled nature. Similarly, even animate beings like snails are in some communities included in this application of hammer-and-anvil technique. On the other hand, other food items that are much closer in their nourishing value to nuts, like soft fruits, are left aside and not treated under the same "label" of stone use.

Hence, the generalizations of tool use remain group-specific for two entirely different reasons. One was already given above. Transfers to novel target areas (e.g., nut species) are only invented by some groups, not others. As a result, the patterns of application remain specific for each tool-using community. Another reason is that different communities rely on different *tool materials* to solve problems in particular domains of action. This fact greatly channels any patterns of transfer performed within a community over successive generations. Different raw materials may be recruited to perform very similar functions. For example, either leaf or a piece of bark or a small stick may become conventionalized as tools for gathering insects off the ground. However, the choice of tool in this particular domain determines possible subsequent patterns of transfer to other domains. For instance, the stick may also serve as fishing pole at beehives, whereas the piece of bark may not. On the other hand, the piece of bark may serve as scooping device for drinking water, while the stick may not. In this way, the tools themselves, that is, their material and historical origins, also play a vital role in explaining how certain functional extensions are possible and why in some cases quite different functional domains (e.g., fishing for insects vs. scooping water) are connected by a similar type of tool use.

Prediction 4

Reciprocal markings are likely to show multiple functions that extend beyond the reciprocal domain. Elements identified as reciprocal markers will have different historical origins and therefore possess language-specific structural and functional features. Consequently, they will differ from one another in respect to their potentials for transfer to other functional meanings that are nonreciprocal.

To summarize, prelinguistic forms of tool use among great apes, such as "nut-cracking" behavior among wild chimpanzees, are considered forms of *cultural behavior*, which are the cumulative result of historical developments within local communities. The historical origin of such cultural behaviors lies in single innovations and subsequent social transmissions within and across the generations. As predicted by the cultural hypothesis, the defining feature of such tool-using behaviors is their *geographic variation*, which manifests itself on various levels of analysis: (1) form of behavior (technique),

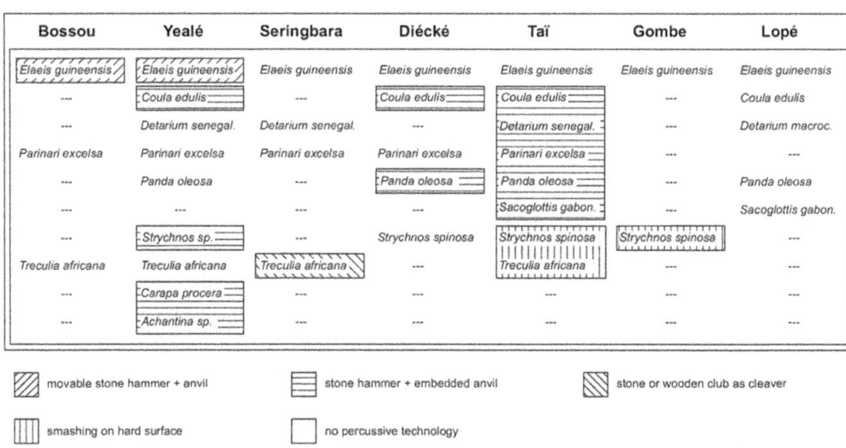

Figure 3.3 Patterns of Variation in Dedicated Nut-Cracking Behavior Among Wild Chimpanzees. The schematic representation summarizes and compares the various findings on nut-cracking behavior recorded in seven different chimpanzee communities living in different parts of Africa (Bossou, Yealé, Seringbara, Diécké, Taï, Gombe, Lopé). The entries in each column represent the findings for each single community. Italics represent the names of nut-/fruit-bearing tree species known to be available within a given group territory. Blanks (---) indicate that the relevant tree species is either absent or at least not known to be present in a given group territory. The shaded frames within each column indicate different forms of dedicated nut-cracking behavior. The nut-/fruit-tree species subsumed under the same shading(s) within columns together represent the full application pattern of a given technique. In cases where the names of nut-/fruit-bearing trees are not subsumed under a shading, it is suggested that chimpanzees from the community do not feed on these nuts/fruits by way of a dedicated technique. *Source*: Author created.

(2) content of behavior (application), (3) the presence or absence of behavior, and (4) the historical origin and functional transfer of behavior. Figure 3.3 summarizes these facts schematically. We can reapply these criteria for cultural variation to the study of language and make concrete predictions with regard to the nature of grammatical constructions in natural languages. In the next chapters, we consider—as a case study—the distributional patterns of reciprocal markings in different languages and test the hypothesis for cultural variation as derived here from the tool hypothesis in application to material cultures among nonhuman primates.

NOTES

1. An anonymous reviewer draws my attention to the works of Karl Marx and his reliance on the notion of "tool" in a literal, non-metaphorical sense of the word. I acknowledge the fact but am not expert enough to associate Marx's notion of tool with the current debate. See also endnote 4 in chapter 9 for a similar but longer acknowledgment of this reviewer's comment.

2. Cf. Köhler (1925 [1976]), *The Mentality of Apes.*

3. All subsequent text passages cited from the German original (Köhler, 1921) were translated by the present author.

4. German "Imitationen des Zufalls."

5. The German original reads: "[. . .] das beste Werkzeug verliert leicht seinen Situationswert, wenn es nicht simultan oder quasisimultan mit der Zielregion gesehen werden kann" (Köhler, 1921, pp. 38–39).

6. Greenfield (1991), Gibson (1993), Byrne and Russon (1998, pp. 673, 675), Corballis (2011) and others explicitly note that the theoretical characterization of goal-oriented behavior in the domain of object manipulation and tool use is analogous to the characterization of sentence structure in the domain of language use (linguistics).

7. In more recent years, long-term studies on Borneo and Sumatra have also revealed similar results with respect to social and tool-using behavior in wild orangutans (*Pongo spec.*); for an overview, see Van Schaik et al. (2003).

8. For a critical assessment of such nativist assumptions, see the discussion in chapter 2.

9. Yet, these chimpanzees do forage extensively on other parts of the oil-palm tree (*Elaeis guineensis*), such as the fruit, petiole/pith, flower, and resin (Wrangham, 1975; Humle & Matsuzawa, 2004).

10. All the chimpanzee communities living in West Africa belong to the same subspecies (*Pan troglodytes verus*).

11. Note, however, that new evidence has been gathered since Boesch et al. (1994). As is pointed out in Biro (2011, p. 166), "a recent report (Morgan and Abwe 2006) on nut-cracking in a community lying outside the boundary proposed by Boesch et al. (1994) now hints toward the possibility of multiple origins for the behavior and calls

for a reassessment of both innovation and extinction of skills as phenomena contributing to observed patterns in chimpanzee cultures (Wrangham 2006)." While this new evidence might lead to a partial reassessment of Boesch et al.'s previous data set, it need not necessarily threaten their basic claim: that nut-cracking behavior in West-West Africa arose much earlier and independently from the possible occurrences of such behavior east of the Nzo-Sassandra River divide.

12. See the "innovation-and-diffusion hypothesis" as presented in Van Schaik et al. (2003).

Chapter 4

Reciprocity in Language (and Beyond)

In order to test our tool hypothesis, we formulated several predictions at the end of chapter 3. These predictions, Predictions 1–4, each specify patterns of variation that we expect to find in cross-linguistic comparison. In the present study, we focus on formally dedicated expressions of reciprocity as defined below.

RECIPROCAL CONSTRUCTIONS

Reciprocity as a complex grammatical category in language has been studied extensively by linguists in both language-specific and cross-linguistic perspective. Originally, as with many other functional categories in language use (Lucy, 1996b), a major focus lay on European languages. Over the years, however, attention shifted toward other language families including, for example, Indian, African, South American, and Turkic languages, as well as the different language families spoken in the Pacific region. Today, comparative language data on reciprocals is available for at least several hundred languages spoken around the world (Frajzyngier & Curl, 1999; Nedjalkov, 2007a; König & Gast, 2008a; Evans et al., 2011a; Maslova & Nedjalkov, 2013).

Reciprocal constructions are complex grammatical means (constructions) for the expression of symmetrical/mutual relations between two or more participants of a situation. Despite their comparatively rare occurrence in natural language use (Evans, 2007), it is possible to identify such constructions in many (if not most) languages across the world. They thereby constitute a pervasive structural phenomenon across languages. In contrast to nonspecialized means (lexical, other) of expressing reciprocity, reciprocal constructions are grammatically (morphologically, syntactically) dedicated forms of language use, which can thereby be clearly demarcated both semantically and formally from nonreciprocal construction types (Haspelmath, 2007b; Evans, 2008). In English, for example, the *each other*-construction is used specifically by speakers in order to express reciprocal event types[1]:

(1) ENGLISH
   a. *John and Bill hit* ***each other.***
   b. *We help* ***each other.***
   c. *The planes flew past* ***each other.***
   d. *The chairs stand next to* ***each other.***

Semantically, these single-clause constructions are equivalent to multiclausal constructions of the type 'A Verb B, (and) B Verb A.' However, formally, they differ in that the plurality of participants are only encoded in a single argument position (subject) and that instead of an ordinary object phrase there is a bipartite expression. By comparison, simple transitive constructions (which lack the employment of *each other* or *one another*) are used in order to express nonreciprocal (unidirectional) situation types. For example, compare utterances like *John hit Bill* and *We help them* with examples (1a-b). In addition to the lack of a reciprocal marker (*each other*) such utterances differ from the canonical reciprocal construction by encoding the participants in separate argument positions like subject and object, respectively.[2]

This contrastive paradigm between unidirectional and bidirectional (mutual/symmetrical) construction types, as illustrated above, is taken as the key diagnostics of dedicated reciprocal constructions in a language. Throughout the discussion below (chapters 5, 6, 7, 8), therefore, the identification of reciprocals in any single language relies on this basic diagnostic test. As we will see, by this standard, it is possible for some languages not to possess formally dedicated means for the expression of reciprocity. This has important implications for our theoretical predictions as derived from the tool hypothesis.

THE RELATIONSHIP TO SOCIAL RECIPROCITY

Opposite linguistic reciprocity (= reciprocal constructions) stands a nonlinguistic concept of *social reciprocity* which is discussed in different fields of behavioral and social science (e.g., behavioral biology, ethology, anthropology, sociology). We here often find it associated with such labels as *reciprocal altruism*, *social exchange*, *cooperation* or simply *reciprocity* (Mauss, 1990 [1924]; Trivers, 1971; Axelrod, 1984; De Waal, 2005a, 2005b;

Tomasello, 1999; Stegbauer, 2010 [2002]). In this case, the focus lies on socially cooperative behavioral strategies between non-kin individuals, which are based on reciprocally altruistic transfers of favors ("tit-for-tat" principle) and the maximization of individual gain and loss accounts. Among nonhuman primates, in particular great apes, experience-based forms of cooperation become increasingly sophisticated, to the extent that the members of a group are able to flexibly cooperate with one another in many different areas of life and under variable external conditions (e.g., foraging, grooming, hunting, conflict resolution) (De Waal, 1997, 2005a, 2005b). As is investigated in the fields of anthropology and sociology, this biologically based capacity for mutual cooperation in humans is increasingly embedded into sociocultural customs within the community, which promote cooperative interactions in almost every area of life. Nonetheless, they are still based on the same conditional (maximizing) principle as in the natural (noncultural) condition (Mauss, 1990 [1924]; Axelrod, 1984; Stegbauer, 2010).

In this context, with regard to language, some linguists have come to discern a possible connection between linguistic and nonlinguistic reciprocity (König, 2005, 2011; König & Gast, 2008b; Evans et al., 2011b; Majid et al., 2011a, 2011b). A conspicuous observation is indeed that reciprocal constructions are most often used with reference to *social interactions among humans* (e.g., mutual hitting, fighting, talking, seeing, greeting, helping, supporting, giving, deceiving). By contrast, the expression of nonhuman, nonsocial interactions between participants (e.g., inanimate objects) is much less common in general statistics of language use. As a result, the "semantic core" of reciprocal constructions in natural language use seems somewhat more closely associated with what behavioral scientists and sociologists identify as the "behavioral (thematic) core" of human social reciprocity. Consequently, it seems at least conceivable to speculate about the idea that the semantic core of reciprocal constructions might stand in a cognitively motivated causal relationship to the human disposition to engage in various forms of social exchange. Reciprocals in natural languages might be "grammars of exchange" (Majid et al., 2011a). We return to this hypothesis in chapter 8. Before, we turn to the testing of our Predictions 1–4 (chapters 5, 6, 7).

THE LANGUAGE DATA

Map 4.1 provides an overview of the languages that were selected for our discussion in the subsequent chapters (chapters 5, 6, 7, and 8). The full list of languages mentioned throughout the discussion is even longer, but the map indicates those sixteen languages that figure most prominently in our

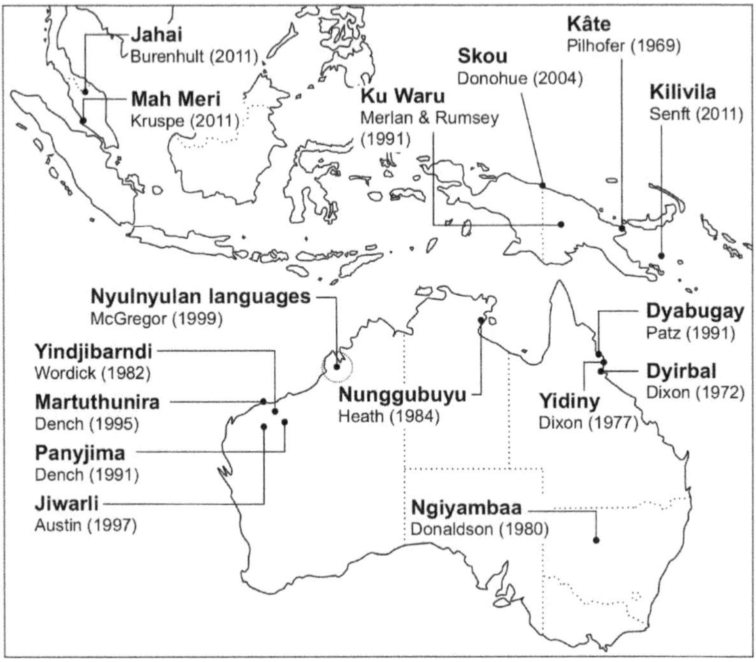

Map 4.1 Minority Languages in Aboriginal Australia and the Pacific Region. The sixteen languages located in this map figure most prominently in the discussion below. We will refer back to this map for geographic reference throughout chapters 5, 6, 7, and 8. *Source*: Author created.

discussion on structural and functional variation, family resemblance, and social reciprocity. Strictly speaking, one entry in Map 4.1 ("Nyulnyulan languages" in Northwestern Australia) does not refer to a single language but a whole language group. This collective entry owes to the literature from which the data is drawn (McGregor, 1999) and the way it is presented and discussed there (collectively). For more information on the languages, source materials and references, the interested reader is referred to the supplementary material in Appendix A.

The reasons for selecting these different languages were threefold. First, it was intended that the language-typological study of reciprocity should not be based on languages from the already well-studied European, Asian, and Pacific (Austronesian) region, but instead on minority languages spoken in regions so far neglected in cross-linguistic comparisons. In the language-typological literature on reciprocals, much attention had been paid to European, Caucasian, Southeast Asian, African, and Pacific languages (Frajzyngier & Curl, 1999; Lust et al., 2000; Nedjalkov, 2007a; König & Gast, 2008a). Australian Aboriginal and Papuan languages, by contrast, had not been considered extensively. The current selection of languages draws on and complements more recent contributions in the literature on reciprocals, which also specifically focus on members of these two language groups (Evans et al., 2007; Evans et al., 2011a).

Second, it was deemed important that only spoken language data from nonliterate speech communities should be included in the sample. A focus lay on languages which are spoken in small hunter-gatherer societies with no more than a few hundred or thousand speakers.[3] It seemed easier and clearer to delineate, characterize, and compare reciprocal marking patterns across small speech communities living under broadly similar sociocultural and ecological conditions. More specifically, the idea was that, given their similar sociocultural organization and ecological modes of living, speakers of different languages would show a general tendency to talk about similar things in everyday life. On this basis, it would be easier to track and compare similarities and differences in the employment of reciprocal constructions across languages.

Finally, a third reason, which in a way harks back to the previous consideration, was that the data collected for each single language should be based on similar methods of primary language documentation and description. Field linguists who describe individual languages may do so on the basis of *direct elicitation* (tête-a-tête interviews between linguist and language informant) and the recording of *natural discourse* (everyday language; spoken narratives on mythical, ethnographic, or other topics). Ideally, a descriptive analysis of a language is based on both these methods. As the current investigation (in terms of sampling method) relied on such primary linguistic descriptions, it was important to select languages for which there would be similar primary sources available (grammars, dictionaries, spoken narrative transcripts). Primary publications in the field of Pacific linguistics are extremely balanced in their documentation and consideration of both elicited and non elicited language material, and therefore seemed to provide an excellent platform for the cross-linguistic investigation of reciprocity.

THE LANGUAGE-TYPOLOGICAL LITERATURE

The study of reciprocal constructions across languages received much attention during the late 1990s and early 2000s. Early collaborative efforts like Frajzyngier and Curl (1999) focused on cross-linguistic investigations of reciprocal marking patterns that often (though not always) overlapped with other functional categories such as the expression of reflexive ("self-directed") and collective ("joint action") meanings, respectively. Even before that, Kemmer (1993) had provided a comprehensive cross-linguistic analysis of "middle voice" constructions that in some languages included the marking of reciprocity. Like most studies in the typological literature, this research was carried out within the functionalist framework. The to date most comprehensive reference work and formal typology on reciprocal constructions remains Nedjalkov (2007a), which encompasses five volumes of essays on reciprocal constructions in individual languages. Its contributors are experts on each individual language and their analyses are based on a shared questionnaire, whose final version is presented and discussed in Nedjalkov and Geniušienė (2007).

Presumably inspired by some of these earlier works as well as general trends in research funding possibilities across the globe (e.g., the DOBES initiative 2000–2011 by the Volkswagen Foundation), other collaborative efforts followed throughout the early 2000s. One of these projects, "A Typology of Reciprocal Markers" (2005–2009), was carried out in collaboration between generative linguists at the Utrecht Institute of Linguistics OTS (M. Everaert) and functionalist linguists at the Freie Universität Berlin (E. König), and focused particularly on the formal encoding of reciprocal meanings across a wide range of languages around the world (König & Gast, 2008a). Another large project, "Reciprocals Across Languages" (2003–2005), was carried out in collaboration between the University of Melbourne (N. Evans) and the Max Planck Institute in Nijmegen (S. C. Levinson). In addition to the consideration of formal diversity, this project developed an experimental methodology designed to test the "semantic scope" of reciprocal event configurations. In a joint effort, typologists and field linguists developed standardized videoclip material (sixty video clips) and subsequently asked experts to test these on mostly minority languages spoken in Aboriginal Australia and the Pacific region. The results were published in Evans et al. (2011a). In addition to the exploration of semantic and formal diversity, some of these later projects also broadened the scope of inquiry by addressing questions with regard to linguistic relativity and the relationship between cross-linguistic regularities and their possible origins in either cultural values or universal dispositions of social cognition and behavior (König, 2005, 2011; König & Gast, 2008b; Wierzbicka, 2009; Evans et al., 2011b; Majid et al., 2011a, 2011b).

The present study owes much to these earlier works and in particular the language-typological framework that has been established with regard to the analysis of reciprocal constructions in cross-linguistic perspective (Maslova & Nedjalkov, 2005, 2013; König & Kokutani, 2006; Nedjalkov, 2007b; Haspelmath, 2007a, 2007b; Evans, 2008). Many of these formal and semantic analyses go far beyond the present work in terms of depth and detail. It is not the aim of this study to match these functionalist-typological analyses. Rather, its aim is to yield positive or negative evidence of the tool hypothesis and its predictions developed in chapter 3. Most of the data discussed in the ensuing chapters (5–8) is based on the author's own sampling of data from independent descriptive works (grammars, dictionaries, spoken text collections) on individual languages. In many places, we also draw on data from the above-mentioned typological literature in order to enrich and generalize the discussion of reciprocal marking patterns across languages. In this way, the current investigation draws on the same basic terminological distinctions introduced in the functionalist literature and brings them together with a new field of inquiry that bridges the gap between linguistic analysis and theory of learning.

NOTES

1. A note on the notational convention for presenting the language data: Throughout the ensuing chapters (5, 6, 7, and 8), we will refer to documented language data in numbered examples. Each example will (1) be headed by the name of the language from which the data was drawn; (2) contain references of the work (author, year, page, or text passage) from which the data was drawn; often this is based on original fieldwork and text collections presented in the works cited; (3) contain glossing (interlinearization) according to the notational conventions of the original (cited) reference; following the Leipzig Glossing Rules, all abbreviations are given in the frontmatter; (4) contain English translations given in single quotation marks ('') at the end of each example.

2. In some languages, like German (and many others), reciprocal markings also occur in unidirectional, transitive constructions, that is, with singular subject and non-subject NPs. Such construction types are commonly referred to as *discontinuous reciprocal constructions* (Nedjalkov, 2007b, pp. 27–28). According to some scholars, their usage differs from canonical or *simple reciprocal constructions* in that they imply a higher degree of perspective-taking in favor of one of the participants (the one encoded in subject position) over the other. Moreover, in cross-linguistic perspective, one finds: "If a language has a discontinuous reciprocal construction, it also has a simple reciprocal construction" (Maslova, 2007a, p. 337). For the sake of simplicity, that is, in order to reduce the empirical load to be taken into consideration here, we do not include this construction type into our discussion below. We stick to the "canonical" or "simple" reciprocal construction as defined here in the main text.

3. Especially some of the Australian languages in this study are greatly endangered, or have already gone extinct.

Chapter 5

Structural Variation

Prediction 1 (see chapter 3)

Reciprocal marking in natural languages differs in cross-linguistic perspective. Natural languages employ different structural, morphosyntactic means to encode reciprocal meanings (situation types). Such variation is unpredictable across languages in the sense that it is not possible to predict what type of syntactic and/or morphological marking will be habitually employed in any chosen speech community. Moreover, even where languages employ similar means, that is, construction types, the forms still differ in detail and do not conform to a single, cross-linguistically valid rule of grammaticality. It is further predicted that languages are likely to employ a broad spectrum of constructional means in reference to reciprocity. Some of these constructions cover the core domain of what is generally defined as reciprocal meaning, while others only overlap with this functional domain at its periphery.

Prediction 3 (see chapter 3)

Not all languages have dedicated reciprocal constructions. Instead, some rely on more indirect means and suggestive context settings in communication, to express the reciprocal relation. It is further predicted that languages differ with respect to the degree to which their reciprocal markings are generalized. Some languages make reference to certain reciprocal meanings without dedicated means; others make the same references relying on such more dedicated means. Even in such cases, however, usage-based factors can be invoked to explain such broad extensions of dedicated (structurally more costly) means.

Natural languages employ a broad spectrum of structural means to express reciprocal meanings. In the following, we exemplify various construction types used in different languages.

NUNGGUBUYU AND OTHER LANGUAGES

Nunggubuyu is an Australian Aboriginal language spoken in southeast Arnhem Land, Northern Territory, in the coastal areas opposite Groote Eylandt (see Map 4.1, p. 68). The language belongs to the non-Pama-Nyungan subbranch of the Australian language family (Dixon, 1980, p. 263; Lynch, 1998, pp. 39, 68; see also Dixon, 2002, pp. 48–49, p. 53, for critical comments on this arguably genealogical division "Pama-Nyungan"/"non-Pama-Nyungan") and within non-Pama-Nyungan is classified as belonging to the Gunwinyguan language family (Harvey, 2003, pp. 204–5; Alpher et al., 2003, pp. 308–9). It was first analyzed and documented comprehensively by Jeffrey Heath who began to work on the language in 1973. At the time when Heath conducted his fieldwork (1973–1977), there were approximately 200–300 people who spoke the language, however with varying degrees of proficiency. The compiled work of grammar (Heath, 1984), dictionary (Heath, 1981), and texts (Heath, 1980) is based on the native language skills of at least ten Nunggubuyu informants. The one reciprocal marker identified as formally dedicated means of expressing reciprocal meanings in Nunggubuyu is the verbal derivational suffix marker *-nyji* (Heath, 1984, pp. 391–93). "As our label suggests, Recip forms often indicate *reciprocal transitive activity*, as in /=na-nyji-/ 'to see each other,' /=r̠i-nyji-/ 'to spear each other' (/=r̠a-/), /=adhuga-nyji-/ 'to pass by each other.' In this sense, two or more agents each engage in the action, and they are also the objects of the component actions (X sees Y, Y sees X)" (Heath, 1984, p. 391). In the examples (2)–(4), the a-sentences illustrate the transitive (nonreciprocal) uses of transitive verb stems, while the respective b-sentences exemplify the derived counterparts involving reciprocal *-nyji*[1]:

(2) NUNGGUBUYU

a. *aba ngi-ga:-'yung yuwa-ga:-'la ngunu=na-ni*
then as for her from there 3FSg.3MSg=see-Pst
'Then she (Jabiru) saw him from where she was.' (Heath, 1980, T10.14.1)

b. *araga nga wini=na-**nyji**-ny*
suddenly and then 3MDu=see-Recip-Pst
'Suddenly they (King Brown and Water Python) saw each other.' (Heath, 1980, T21.6.1)

(3) NUNGGUBUYU

a. *wunu=r̠a-ni*
3Pl.3MSg=spear-Pst
'They speared him (the man with bad eyesight).' (Heath, 1980, T69.9.2)

b. *wu:=ṟi-**nyji:**-ni*
3Pl=spear-Recip-Pst
'They (the old people) speared each other.' (Heath, 1980, T69.10.2)

(4) NUNGGUBUYU

a. *nu=wagaṟa-ngi* *nu:-'ba-gi-yung*
3MSg.3MSg=head.for-Pst that (MSg)
'He (Wuruyambi) went toward him (Maṉnyamaṉnyiri).' (Heath, 1980, T15.4.2)

b. *wu:=wagari-**nyji:**-ni* *wu:=yama:-'*
3Pl=head.for-Recip-Pst 3Pl=do-Pst
'They headed for each other (came together) like that.' (Heath, 1980, T47.3.4)

The transitive verb stems marked by reciprocal *-nyji* are formally *intransitive*, a phenomenon often noted for reciprocal markings in languages around the world (Frajzyngier & Curl, 1999, p. x; Evans, 2008, p. 69; Maslova, 2008, pp. 229–30, p. 233). Thus, in the b-sentences above, the pronominal prefixes on each verb stem (*wini-* and *wu:-*, respectively) belong to the intransitive prefix set of Nunggubuyu (Heath, 1984, tab. 9-1, p. 348). By contrast, in the corresponding a-sentences, the pronominal prefixes (*ngunu-, wunu-, nu-*) are chosen from the transitive prefix set of the language (Heath, 1984, tab. 9-2–9-15, pp. 349–59). The subject referents in reciprocal utterances are therefore formally treated the same way as subjects in nonreciprocal utterances that lack the overt expression of object referents. This has the interesting effect that reciprocal event types are linguistically conceptualized as integrated, undifferentiated wholes rather than separate, individuated actions or states. This phenomenon of formally reduced linguistic expression is reminiscent of what Kemmer (1993) describes in respect to the encoding of motion events by syntactically intransitive verbs: "The class in which there is least differentiation of participants is that of unmarked intransitive verbs. In this case the affected participant is not even coded—the action is conceived of as being carried out by a completely unitary conceptual entity" (Kemmer, 1993, p. 69). Kemmer usefully refers to this feature of linguistic constructions as *the relative degree of elaboration of events*.

> We can define the relative elaboration of events as the degree to which the facets in a particular situation, i.e. the participants and conceivable component subevents in the situation, are distinguished. The variation in elaboration of events reflects alternative conceptualizations by the speaker: The speaker has the choice of either making reference to events as undifferentiated wholes, or making reference to their substructures or component parts. (Kemmer, 1993, p. 208)

In the case of reciprocal constructions, especially when syntactically marked as intransitive, the degree of elaboration of events is thus relatively low, a point we return to in chapter 8 (in the sections on "Reciprocal Schemas" and "Granularity of Event Description").

To sum up the data on Nunggubuyu, the verbal suffix *-nyji* is a highly productive reciprocal marker. It is productive with at least several hundred transitive verb forms and, with only a few exceptions, is used to express reciprocal actions among a plurality of participants. Figure 5.1 summarizes this fact about the productivity of reciprocal *-nyji* schematically.

In the Australian languages Dyirbal, Jiwarli, Ngiyambaa, and Duuŋidjawu, reciprocal markers are likewise classified as "verbal suffixes," that is, elements attached at the end of a verb stem or root (Dixon, 1972; Austin, 1997; Donaldson, 1980; Kite & Wurm, 2004). In others, like the Papuan language Yimas or the Oceanic language Fijian, we are dealing with "verbal prefixes" as reciprocal markers, that is, elements attached at the beginning of a verb stem or root (Foley, 1991; Dixon, 1988). The typological literature at large provides ample evidence that such verbal marking patterns are common

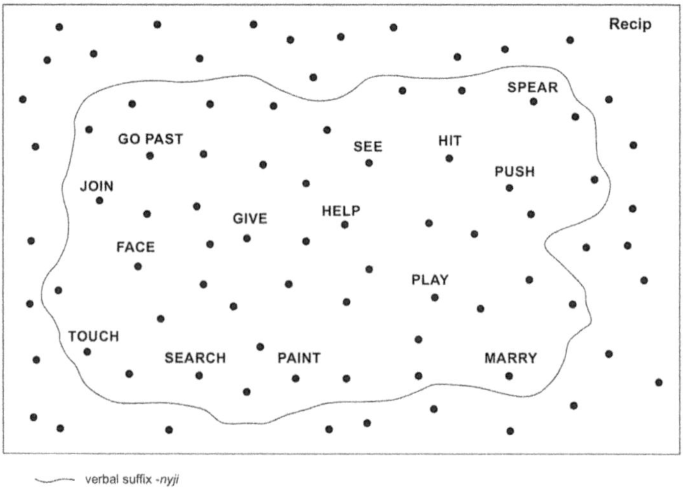

Figure 5.1 Reciprocal Marking in Nunggubuyu. In this language, the verbal suffix *-nyji* is highly productive as a marker of reciprocity. The black frame labeled "Recip" with black dots (•) represents the "potential space" of reciprocal marking. The black dots stand for lexical verbs that may or may not be marked by the suffix. In cases where dots remain outside of the lined shading, it is indicated that some verbs will not be marked by the suffix, simply because some formations will not yield sensible reciprocal meanings (e.g., *build-Recip 'build one another'; *grow-Recip 'grow one another'). In cases where dots lie within the lined shading, it is indicated that the verbal suffix is productively used on verb stems to express reciprocal meanings. A full list of such reciprocal verb formations is given in Tables 11.1.1–11.1.6, Appendix B. *Source*: Author created.

strategies for the expression of reciprocal meanings (Maslova & Nedjalkov, 2005, 2013; König & Kokutani, 2006; Evans, 2008), even though the morphosyntactic details of such constructions tend to vary even within language families (e.g., Evans et al., 2007).

In other languages, reciprocal marking consists of both verbal prefixation and suffixation at the same time (circumfixation). In the case of Nyulnyulan languages (Northwestern Australia, see Map 4.1, p. 68), this is particularly interesting. Here, the same verbal suffix (*-nyji*), which we also find as reciprocal marker in Nunggubuyu (see above), is combined with a verbal prefix (*ma-*) in order to express reciprocity.[2] Thus, in the examples (5)–(6) from different Nyulnyulan languages, the circumfix *ma-V-nyji*, not merely suffixation with *-nyji*, is the standard way of marking reciprocal action:

(5) WARRWA

*Ngi-rr-**ma**-yama-**nyji**-ny-bili.*
3:Nom-Aug-RR-argue-RR-Pst-Du
'They two argued with each other.' (McGregor, 1999, p. 89)

(6) NYIKINA

*Yirrka yi-rr-**ma**-kama-**nyji**-n-mirri.*
3:Aug 3:Nom-Aug-RR-laugh-RR-Pres-Aug
'They are mocking one another.' (Stokes, 1982, p. 289)

Such small-scale cross-linguistic differences demonstrate that there exists no intrinsic mechanism that connects form with function in language use. The use of a verbal suffix marker in a language, and its association with a certain function is not fixed or predetermined. That which may be grammatical and functionally dedicated in one language may be ungrammatical in another. Typologists also refer to this fact as the *relationality* of forms (and their respective functions) in language use (König & Gast, 2008b, p. 16). Whether or not a linguistic form can be used for a particular speech performance (guided by a speaker's intention), depends on various context factors specific to each individual language. For instance, the *polysemy* of a linguistic form, that is, its potential to be used in more than just one function in similar contexts, may trigger the addition of further expressive (disambiguating) means and their subsequent conventionalization as a way to retain clear reference. Moreover, languages may undergo radical changes in argument structure (e.g., with regard to case marking or the distinction between transitive and intransitive clause types). Such changes are not directly linked to the expression of reciprocity; however, they may influence how this type of meaning is expressed morphosyntactically over time.

We return to such considerations in more detail in chapter 7 (see section on "Structural Context Factors").

KÂTE AND OTHER LANGUAGES

In some languages, structural means other than "verbal affixes" are employed. Furthermore, some languages possess more than one type of reciprocal marking pattern with various degrees of productivity.

Kâte is a Papuan language currently spoken by approximately 20,000 people living in the Finschhafen District of Morobe Province, Papua New Guinea (Simons & Fennig, 2017) (see Map 4.1, p. 68). According to Ross (2005) and McElhanon (1975), it belongs to the Finisterre-Huon subgroup/family of the hypothetical Trans-New Guinea phylum of Papuan languages. Foley (1986) classifies it as Eastern Huon together with languages like Mape and Kube, and together with Western Huon languages like Selepet relates it to the Central Highlands languages including, for example, Fore, Awa, and Dani (Foley, 1986, pp. 243–44; pp. 261–62). In the early 1900s, the German missionaries Georg Pilhofer and Christian Keyßer compiled their descriptive study on the language, based on an estimated number of ca. 600 native speakers living in several small villages (Keyßer, 1969 [1925]; Pilhofer, 1969 [1933]).

In this language, the expression of reciprocity involves a form of reduplication (semi-reduplication) as a means of deriving reciprocal word forms (verbs, nouns, adverbs) from nonreciprocal ones.[3] Most productive is a complex verb construction involving a (semi-)reduplicated verb form and a finite form of the basic verb *e-* 'be, say, do' (see example [7b]).

(7) KÂTE

a. *hesuec-nu-zo*
push-1Sg-*zo*
'to push me away, reject me' (Flierl & Strauss, 1977, p. 136)

b. ***he****-hesuec* ***e****-ŋgopieŋ*
Rdp-push do-Pres.2/3Pl
'they are pushing each other away' (Flierl & Strauss, 1977, p. 130)

In addition, there is another type of predicate construction that also receives this kind of (semi-)reduplicated marking. Following the literature on Papuan languages (Foley, 1986, pp. 117–22; Lynch, 1998, chapter 7), we can label this as *adjunct verb construction*. This construction, as opposed to *simple verb constructions*, consists of morphologically invariant expressions (typically nouns) in combination with simple inflecting verbs, which together function as verbal predicates of a sentence: predicate = adjunct + inflecting verb. In Kâte, such adjunct constructions are fairly frequent. Some examples of possible adjuncts are expressions like *dâŋ* 'voice, sound, speech,' *pele* 'quarrel, fight,' *soŋaŋ* 'guard,' and *micko* 'in the mouth.' Some inflecting verbs combining with such expressions are verbs like *e-* 'say, do,' *mu-* 'say, speak,' *ba-* 'hold, do,' *qa-* 'hit,' and *lo-* 'take.' Many different adjuncts combine with one or several from this small set of inflecting verbs in order to express concrete lexical meanings: for example, *dâŋ muzo* 'to scold sb.,' *dâŋ bazo* 'to talk to sb.,' *pele ezo* 'to scold sb.,' *soŋaŋ bazo* 'to take care of sb.,' or *micko bazo* 'to answer/gainsay sb.' With these adjunct predicates, the reciprocal constructions are formed in essentially the same way as in the case of simple verb constructions exemplified above: the simple inflecting verb form is reduplicated and used in construction with a finite form of *e-*; the noun adjunct remains unaltered (see example [8b]).

(8) KÂTE

a. *pele* *e-nale-kac*
scolding do-1Sg-Pres.3Sg
'he scolds me' (Keyßer, 1969, p. 330)

b. *pele* ***e**-ec* ***e**-kopilec*
scolding Rdp-do do-Pres.2/3Du
'the two scold each other' (Keyßer, 1969, p. 330)

However, there is a small class of verbs in Kâte which behave differently (Pilhofer, 1969, p. 99). They are referred to as *object verbs*. To form a reciprocal construction, these object verbs do not follow the prevalent pattern of verb stem (semi-)reduplication. Instead, the verb stem for "first person singular" and "second person singular," respectively, are put together to form a composite verb form. Just as in the case of (semi-) reduplicated verb stems, these composite verb forms do not carry any finite marking for categories like tense, aspect or person-number of the subject. Instead, finite marking again only occurs on an inflected form of *e-*, which always follows the composite verb form (see example [9] for the verb *nare-* 'to give me sth.').

(9) KÂTE

nareŋ *gareŋ* *e-*
give.1Sg give.2Sg do-
'to give each other, to exchange, to barter' (Keyßer, 1969, p. 293)

Thus, instead of verbal (semi-)reduplication, for example, **nanareŋ e-* or **gagareŋ e-* in (9), we get verbal compositions of two different verb stem forms: that is, *nareŋ gareŋ e-* 'to give each other.' In total, there are eight of these object verbs in Kâte and all of them except for *qa-* 'hit' (another exception) rely on verb composition in combination with finite *e-* as a way to express the reciprocal meaning (Pilhofer, 1969, p. 99 ff.; Keyßer, 1969). This overall situation of reciprocal marking in Kâte is summarized schematically in Figure 5.2.

The situation found in Kâte can be further contrasted with structural differences found in other languages. In this context, we return to the Nyulnyulan language group spoken in Western Australia (see Map 4.1, p. 68). Earlier, we characterized their reciprocal marking pattern in terms of verbal circumfixation with *ma-V-nyji*. This is true of reciprocal marking on simple inflecting verbs (e.g., Nyulnyul *-jal-* 'see,' Warrwa *-murungu-* 'look for,' Yawuru *-bilka-* 'hit,' Bardi *-ugul-* 'break'). However, the formal marking is quite different with so-called preverbs. Like adjuncts in Kâte, preverbs are morphologically invariant forms that only carry lexical meaning but do not inflect for grammatical categories like tense, aspect, mood, subject/object

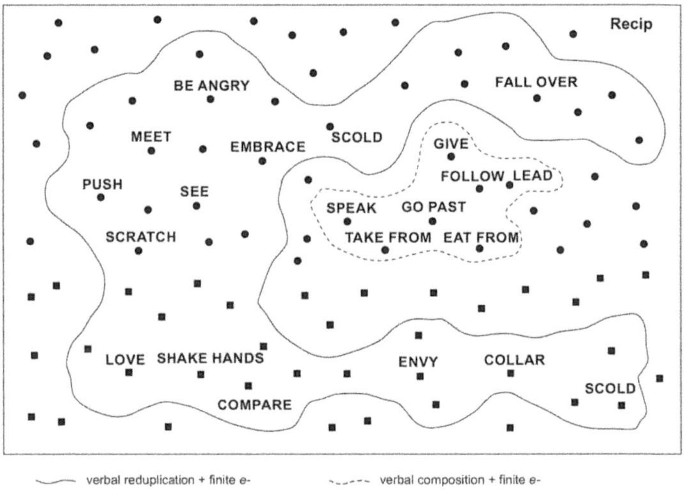

Figure 5.2 Reciprocal Marking in Kâte. The diagram schematically depicts the distributions of reciprocal marking patterns with Kâte verbs (black dots and squares). Verbal reduplication followed by finite forms of e- 'say, do' is the standard way of expressing reciprocal meanings with simple (•) and adjunct (■) verbs, respectively. In addition, there is a small class of verbs (object verbs) with which verbal composition (plus finite forms of e-) is the only grammatical marking pattern. In the diagram, its limited distribution is indicated by the dotted line. All entries are based on the full data set in Tables 11.2.1–11.2.3, Appendix B. *Source*: Author created.

referent, transitivity, and so on (McGregor, 1999, 2002). If a preverb is used in one of the Nyulnyulan languages, it must always be combined with another inflecting verb form, which encodes these derivational and/or inflectional categories for them. For example, Nyulnyul possesses several hundred preverbs, among them forms like *baab* 'open,' *dimbidimb* 'join,' *durr* 'push' and *junk* 'run' (McGregor, 2002, pp. 113–14). When used in verbal utterances, these various preverbs are used in construction with one of several inflecting verb forms, among them *-jid* 'go,' *-ny* 'get,' *-m* 'put,' *-j* 'say, do' and *-w* 'give' (McGregor, 2002, p. 111). The result are so-called compound verb constructions such as the following: *baab -ny* 'open something,' *dimbidimb -j* 'marry,' *durr -w* 'push' or *junk -jid* 'be running along' (McGregor, 2002, pp. 113–14). Essentially, then, compound verb constructions in Nyulnyulan languages are like adjunct verb constructions in Papuan languages like Kâte. The difference in terminology is mainly due to linguistic tradition: Australianists have coined the terms *preverb/coverb* and *compound* (McGregor, 1999; McGregor, 2002, pp. 25–28; Dixon, 2002, pp. 60–63, p. 184), while Papuanists have coined the term *adjunct* to capture this linguistic phenomenon (Foley, 1986, pp. 117–22; Lynch, 1998, chapter 7).

So how do Nyulnyulan languages treat such compound/adjunct constructions for the expression of reciprocity? The answer is, they treat them with a different structural marking pattern. Instead of using circumfixal *ma-V-nyji* on the inflecting verb bases, speakers of Nyulnyulan languages draw on *-banji/-wanji* 'exchange' as an inflecting verb form dedicated to the expression of reciprocal meanings with preverbal elements (see examples [10] and [11]).

(10) YUWURU

Ngaw *inga-rr-**banji**-nda.*
kiss 3:Nom-Aug-RR-Perf
'They kissed one another.' (Hosokawa, 1991, p. 175)

(11) WARRWA

Kinya *kujarra* *kanyjirr* *ngi-rr-**wanji**-ny-bili.*
this two look 3:Nom-Aug-RR-Pst-Du
'They two stared at one another.' (McGregor, 1999, p. 97)

This way of marking reciprocity in compound constructions thus differs strikingly from how Kâte marks reciprocity in adjunct constructions. Kâte sticks to the same marking pattern applied to simple verb constructions involving (semi-)reduplication and the auxiliary *e-* 'do.' In Nyulnyulan languages, on the other hand, the circumfixal pattern is only productive with simple verb constructions; with compound constructions, the simple inflecting verb *-banji/-wanji* is productive as reciprocal marker instead.

A possible motivation for this inability of compound predicates to combine with the reciprocal circumfix *ma-V-nyji* is that the few inflecting verb forms that typically co-occur with lexical preverbs do not, even if used in simple verb constructions, allow to be marked by these two affixes (McGregor, 1999, p. 96). That is, inflecting verbs like *-j(i)* 'say, do,' *-m(a)* 'put,' *-nyu* 'get, catch,' *-n/-nga* 'be,' *-ngara* 'become' or *-jid* 'go' in Nyulnyulan languages do not have a corresponding reciprocal form, even when used on their own in simple verb constructions. Consequently, they also do not allow this marking when used in compound verb constructions. As a result, speakers of those languages have overcome this problem by drawing on another verbal expression as inflecting verb form.

Interesting is also the etymological origin of the inflecting verb *-banji/-wanji*, which is still transparent today. It means 'exchange; give one another' and is thus what is commonly referred to in the typological literature as an inherently reciprocal or *inherently symmetrical verb* (Kemmer, 1993; König & Kokutani, 2006; Nedjalkov, 2007b; Haspelmath, 2007b; Evans, 2008). As McGregor (1999) points out, the same verb form can thus also occur on its own, in the simple verb construction, where it functions as a "lexical reciprocal" (Nedjalkov, 2007b; Nedjalkov & Geniušienė, 2007; Knjazev, 2007),

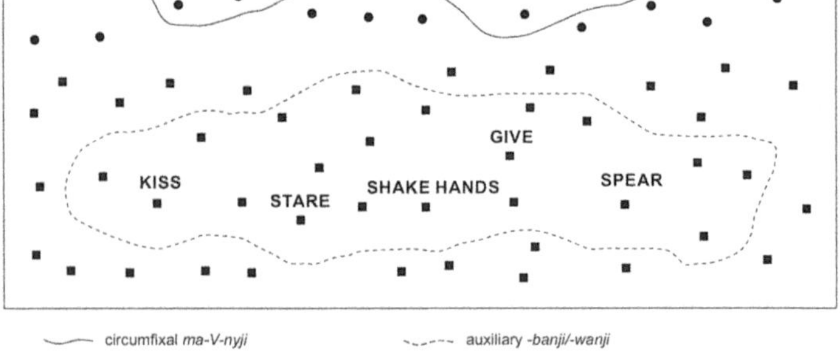

Figure 5.3 Reciprocal Marking in Nyulnyulan Languages. With simple verbs (•), verbal *ma-V-nyji* is the standard way of marking reciprocity; with compound verbs (■), the reciprocal auxiliary *-banji/-wanji* is productive instead. The entries are based on the references and data sets given in Tables 11.3.1–11.3.2, Appendix B. *Source*: Author created.

expressing mutual exchange of goods or favors without overt formal marking by *ma-V-nyji*:

(12) YAWURU

*Yanga-rr-**banji**-n* *kunyjurung.*
1/2:Nom-Aug-exchange-Ipfv gift:ceremony
'We exchange things in the Kunyjurung ceremony.' (Hosokawa, 1991, p. 175)

The situation found in Nyulnyulan languages can be summarized schematically as shown in Figure 5.3. Note its characteristic distributional patterns with respect to (•) and (■), and how they differ from the Papuan language Kâte (see Figure 5.2).

KU WARU AND OTHER LANGUAGES

The absence of a morphosyntactic marker of reciprocity in a language (equivalent, e.g., to English *each other* or Nunggubuyu *-nyji*) must not imply the absence of a *formally dedicated* reciprocal construction. A good example in this context provides the Papuan language Skou (Donohue, 2004), spoken in coastal regions of Northern Papua New Guinea (see Map 4.1, p. 68). In some cases, however, like the Papuan language Ku Waru (Merlan & Rumsey, 1991) (see also Map 4.1, p. 68), it seems more adequate to conclude that at least some discourse areas tend to do without rather with dedicated reciprocal marking (see further below).

In Skou, reciprocal marking does not involve a morphologically distinct marker. Nevertheless, as Donohue (2004) points out, the reciprocal meaning is clearly identifiable through non-singular subject marking on the verb and the lack of an overt object.

> The reciprocal construction in Skou is a variant of the normal structure used to express conjoined nominals in the same NP. The verb is necessarily marked for a non-singular subject, but there is no explicit marker of reciprocity in the clause; only the absence of any realisation of any nominal (or pronominal) object in the clause is a clue to the fact that a normally bivalent verb should be read with a reciprocal meaning. (The fact that, in other circumstances, the overt presence of an object is obligatory, this is then a significant, if negative, indication of the construction.) (Donohue, 2004, p. 413)

Donohue (2004) subsequently draws a more narrow distinction between the treatment of simple inflecting verb forms and compositional (adjunct) predicate forms: the former are formally indistinguishable in their reciprocal use from normal transitive sentences (the only difference being that they do not occur in construction with object phrases within a sentence); the latter, by contrast, are sometimes distinguishable from normal transitive forms by replacing a specific inflecting verb form by the basic form *-li* 'do.' In the following two examples ([13] and [14]), the b-sentences illustrate the dedicated reciprocal construction, while the a-sentences represent the normal transitive constructions. Note, however, that in (13a), the sentence may, given the right context, also serve the reciprocal function. Example (13) exemplifies the use of a simple inflecting verb *-ká* 'hit,' while example (14) illustrates that of the compositional (adjunct) form *pìng -lú* 'bow -release' ('shoot'), which in its reciprocal form reads *pìng -li* 'bow -do.'

(13) Skou

a. *Te=ing a te te=j-á.*
3Pl=the 3Pl 3Pl=3Pl-hit
'They$_i$ hit them$_j$.'
Or: 'They$_i$ hit each other$_i$.'

b. *Te=ing a te=j-á.*
3Pl=the 3Pl=3Pl-hit
'They hit each other.'
*'They hit (someone else).' (Donohue, 2004, pp. 413–14)

(14) Skou

a. *Te=Máwo te Te=Téme pìng te=r-ú.*
3Pl=Skou Mabo 3Pl.Erg 3Pl=Nafri bow 3Pl=3Pl-release.Pl
'The Skou Mabos shot the Nafris.'

b. *Te=Téme Te=Máwo pìng te=ti.*
3Pl=Nafri 3Pl=Skou Mabo bow 3Pl=3Pl.do
'The Nafris and the Mabos shot each other.' (Donohue, 2004, p. 414)

Donohue concludes: "We can see that, as with reflexives, there is no morphology dedicated to reciprocal constructions, but there is nevertheless a distinct reciprocal construction in the language. This reciprocal construction displays behaviour, in terms of the morphology and syntax that may appear in it, that is not found elsewhere, and so it is uniquely defined" (Donohue, 2004, pp. 414–15).

In some languages, however, the formal expression of reciprocal meanings is even more nondescript, and it is tempting to conclude that a language does not possess a dedicated reciprocal construction. Senft (2011), for example, describes the Oceanic language Kilivila, the language of the Trobriand Islanders, which is one of 40 Austronesian languages spoken in the Milne Bay Province of Papua New Guinea. The data he presents in his study were collected by means of a large set of some 60 video clips specifically designed for the elicitation of reciprocal expressions (Evans et al., 2011a). In reference to most video scenarios (allowing reciprocal interpretations) speakers of Kilivila simply used constructions implicating the reciprocal interpretation through context (see examples [15a–b]).

(15) KILIVILA

a. *Vivila tau e-bigatona-si mata-la tau mata-la vivila.*
Girl boy 3-talk-Pl eye-his boy eye-her girl
'A girl (and) a man they talk (with each other), the man (is) eye in eye with the girl.' (Senft, 2011, p. 228)

b. *E-kepapa-si asitetolu.*
3-hug-Pl three.of.them
'They hug (each other) the three of them.' (Senft, 2011, p. 228)

In some cases, Senft argues, it is not clear from the context whether reciprocity is either implicated or simply not expressed at all (because the situations invite other, nonreciprocal interpretations in the first place). Based on his various findings he concludes that "Kilivila is one of the languages of the world that lacks dedicated reciprocal forms" (Senft, 2011, p. 225). Instead, he argues, Trobriand Islanders fall back on the following three strategies: reciprocity is not expressed at all; reciprocity is implicated; reciprocity is expressed periphrastically (Senft, 2011, p. 230).

Another example is Ku Waru, a Papuan language of the Trans-New Guinea family in the Western Highlands of Papua New Guinea (see Map 4.1, p. 68). Ku Waru belongs to a dialect continuum that includes ca. 250,000 speakers (Rumsey, 2010; Rumsey, 2017). Merlan and Rumsey (1991) provide an anthropological as well as linguistic account of the language and its people. The study of reciprocal marking in this language makes two distinct observations possible: on the one hand, there exists a formally distinct construction, involving a reduplicated (emphatic) form of the personal pronoun (e.g., *olyolyo* 'we ourselves'), which is used in some contexts to more explicitly (unambiguously) express reciprocal meanings. Such constructions are occasionally found in natural discourse but less often than direct elicitation tasks suggest. By contrast, a systematic quantitative assessment of the natural discourse (i.e., without direct elicitation) reveals that native speakers of

Ku Waru tend, in the majority of cases, not to use the pronominal construction type. Instead, they rely on complex verb constructions also used in utterances with nonreciprocal interpretations (see examples [16a–c] on p. 88).

Before we turn to the examples from a spoken text transcript in Merlan and Rumsey (1991), let me briefly put this discussion into perspective. The data reviewed here does not in principle deny the existence of explicit reciprocal marking in Ku Waru. But given a large corpus of spoken narrative and discourse that does not display many instances of explicit (emphatic) pronominal marking, it is worth our while to consider the alternative communicative strategies that speakers of Ku Waru use in such contexts of speaking. After all, it is interesting to note that speakers can do without, in some if not most of their daily discourse, explicit formal means in order to express complex meanings, including reciprocity. In some cases, this may simply mean that the reciprocal meaning is not intended, that is, out of focus in what the speakers are trying to convey. In other contexts, however, it may mean that either less overt or less systematic formal means are employed in order to mark the reciprocal meaning. The latter is what is mostly at stake in the analysis below. Furthermore, it is quite conceivable that the special nature of the chosen discourse data does not allow very broad generalizations with regard to Ku Waru grammar. The reason is that at least some of the data presented here are based on *el ung* 'fight talk' which in Ku Waru represents a special genre of speaking at public meetings between neighboring tribes (*talapi*) who try to settle disputes or negotiate trade deals with one another (Merlan & Rumsey, 1991, pp. 98–109). It is used only by adult males and it displays several stylistic features not found in everyday language. These range from prosodic features (altered pitch contours; a so-called *el ung* vowel at the end of single utterances), paralinguistic features (striding, with arms in hand, back and forth while speaking), rhetorical devices (repetitions with slight variation in a single syntactic frame), the use of segmentary (rather than individual) person references, and the frequent reliance on tropes, that is, metaphorical speech known in Ku Waru as *ung eke* 'bent speech' (see more on this below).[4]

Moreover, Merlan and Rumsey (1991) observe that the seeding of ambiguity and vagueness is very central to fight talk in Ku Waru. Relying on stylistic elements that fall into the category of figurative or metaphorical speech (*ung eke* 'bent speech'), speakers intend in many places to keep references to shared history vague and opaque so as to leave more room for interpretation and subsequent negotiation of material benefits.

> [T]he allusiveness so highly valued in *ung eke* inheres in the employment of tropes in ways that can be seen as portentous, particularly for inter-*talapi* [inter-'tribal'] relations. In the most successful uses of tropes there is always an

element of uncertainty concerning how they may be related to circumstances. Orators work at introducing images into events, and thus—whether the tropes are interpreted exactly as they intended or not—contributing to the sense of outcome of those events, rather than at coining new tropes. (Merlan & Rumsey, 1991, p. 109)

Given this background, it is well conceivable that "the need for ambiguity in oratory is suppressing clearer expressions of reciprocity which would be found in more normal speech" (Nick Evans, p.c.). I agree, and I can only caution the reader that this possibility exists. Ideally, one would analyze more natural discourse data on Ku Waru (which is not "fight talk") or subject Ku Waru speakers to standardized tests such as the video clip material and accompanying methodology presented in Evans et al. (2011a). Moreover, the fact that Merlan and Rumsey (1991) were able to elicit more explicit reciprocal marking in face-to-face interviews with their informants shows that more dedicated reciprocal marking exists.

Nevertheless, despite these precautions, I contend that the data can still provide us with some interesting insights. For one thing, not all the data relied upon for this analysis is fully *el ung*; in fact, only a minority of it is. As is detailed by Merlan and Rumsey (1991, p. 99), only 452 lines of the full 1850 in the transcript are in "full" *el ung*, and another 180 are "intermediate" between full *el ung* and non-*el ung*. In other words, "[i]n practice, the distinction between *el ung* and non-*el ung* is not a categorical one" (Merlan & Rumsey, 1991, p. 99). The examples discussed here are not classified as (phonetically) full *el ung* (Merlan & Rumsey, 1991, p. 99). Furthermore, even if only partially non-*el ung*, the data allows us to speculate that there may in principle be languages that can do entirely without either overt or systematic formal expression of reciprocity. In the case of Ku Waru, this is made plausible to some extent through the additional consideration of data from an unpublished word list by Merlan and Rumsey (1989), which provides insights into the conceptual meaning ranges of many lexical items from everyday discourse.

But first, let us turn to the spoken narrative data from Ku Waru "fight talk" (see examples [16a–c]). The full data set was recorded in 1983 at the proceedings of the compensation payment between neighboring tribes called Laulku and Kopia-Kubuka at a place called Kailge. As the text transcript itself only consists of transcriptions and English translations, the interlinearizations (morpheme-by-morpheme glosses) had to be added by the present author, based on Merlan and Rumsey's own grammar sketch (Appendix B) and on Alan Rumsey's support (p.c.).[5]

(16) Ku Waru

a. *olyo ilyi ekepu el-ayl temulu tekimul, ekepu*
1Pl this now fight-Def do:Fut:1Pl do:Prog:1Pl now
el tekimul
fight do:Prog:1Pl
'we will fight about it, now we are fighting' (Merlan & Rumsey, 1991, p. 249, line 90)

b. *el ung ya kopia kubuka-ma para*
fight talk here Kopia Kubuka-Pl back.and.forth
naa nyai-o
Neg talk:Imp:2/3Pl-[*el ung* vowel]
'but don't you Kopia-Kubuka talk (Pl) el ung ['fight talk'] back and forth' (Merlan & Rumsey, 1991, p. 290, line 1094)

c. *mel-nga kapu tepul-kin naa tirimul*
thing-Gen fight.over.things do:NF:1Du-Com Neg give:RPst:1Pl
'When we fought we hadn't given things to each other.' (Merlan & Rumsey, 1991, p. 292, line 1155)

Arguably, and in consideration of other examples referenced in Table 11.4 in Appendix B, sentences like those in (16) are functional equivalents to dedicated reciprocal constructions in other languages. This is not an easy point to make, seeing as the English translations, too, lack overt reciprocal marking—but we will say more on this issue below. Let me first outline why in the sentences above, no systematic formal marking pattern is discernible. The only marking of reciprocity resides in the dual or plural marking on subject NPs as well as (non-)final verb forms, and the consistent absence of ergative case marking, generally characteristic of intransitive sentences. Moreover, there is no overt expression of object NPs denoting human referents. However, none of these features, not even the combination of them, qualifies as truly dedicated reciprocal marking. As Merlan and Rumsey's (1991) discussion of the basic clause structure makes clear, it is in many cases not possible in Ku Waru to *formally* distinguish between semantically directed (transitive) and nondirected (intransitive) utterances. The formal elements characteristic of directed (transitive) meanings (e.g., ergative marking, the overt expression of object phrases) are merely optional devices. "The basic clause types in Ku Waru are not amenable to any such binary classification, because most Ku Waru verbs can occur in construction with either one or two NPs in core syntactic case roles. Nor does the distribution of syntactic case-marked NPs [absolutive, ergative] provide evidence for a binary transitive-intransitive distinction" (Merlan & Rumsey, 1991, p. 337). This fact contrasts with the general observation made earlier by Donohue (2004) with respect to Skou where transitive verbs always require the overt expression of both their arguments within a sentence and are thereby distinguishable from intransitive sentences.

A characteristic feature of many Ku Waru predicates—not specific to the expression of reciprocity—is that they consist of *complex (compositional) verb constructions* (Foley, 1986, pp. 113–28; Lynch, 1998, chapter 7). These consist of a final, fully inflected verb form and one or several elements preceding it (Merlan & Rumsey, 1991, p. 325 ff.). Depending on their form and meaning, these preceding elements can be classified as either nominal adjuncts, adverbial adjuncts, preverbs or as medial, partly inflecting verbs. In the first example above (16a), the verb complex consists of a final (inflected) verb form and a nominal adjunct, the latter being morphologically invariant: *el te-* 'fight do-.' In the second example (16b), the verb complex takes the form of an adjunct construction, in which the final inflecting verb is preceded by an invariant adverbial element: *para nyi-* '[back and forth] talk-.' The third example (16c) provides an instance of a serial verb construction in which a final inflected verb form is preceded by an adjunct + nonfinal verb construction: *kapu te- ti-* '[fight over things] do- give-.'[6]

In an unpublished word list on Ku Waru by Merlan and Rumsey (1989),[7] frequent attestations of reciprocal uses of complex verb forms occur.[8] More specifically, the lexical entries indicate that the various forms are used either reciprocally or nonreciprocally without any change in verbal composition: examples are *dup-si-* 'to tell a story; or: to tell each other stories,' *elkupula te-* 'to push, shove; or: to elbow and shove each other,' *opimi to-* 'to promise; or: to make each other promises,' and *kumulau lyi-* 'to hold, grip; or: to hold/clutch each other.'[9] Some of these forms involve basic verbal elements like *te-* 'do, transact, give' and *to-* 'hit' or *lyi-* 'get' which in other languages are specific to verbal compositions expressing reciprocal meanings (e.g., see above: Skou *-li* 'do'; Kâte *e-* 'do'; Nyulnyulan *banji-/wanji-* 'exchange'). In Ku Waru, however, they are not specific to marking reciprocity. They are employed in verbal compositions irrespective of whether they refer to reciprocal or nonreciprocal event types.

On this basis, it is tempting to conclude that speakers of Ku Waru do not customarily rely on any dedicated means to express reciprocity (see Figure 5.4). At the very least, we may venture such a generalization for "fight talk" (*el ung*) in Ku Waru. Moreover, as the search through the dictionary suggests, this is also true of many lexical items (complex verbs) that are likely to also carry such functional properties in normal (non-*el ung*) speech contexts. In this view, then, speakers of Ku Waru customarily rely on a formally non-explicit (vague) combination of lexical verb complexes, plural or dual subject marking, and the absence of object phrases, to convey the reciprocal meaning.

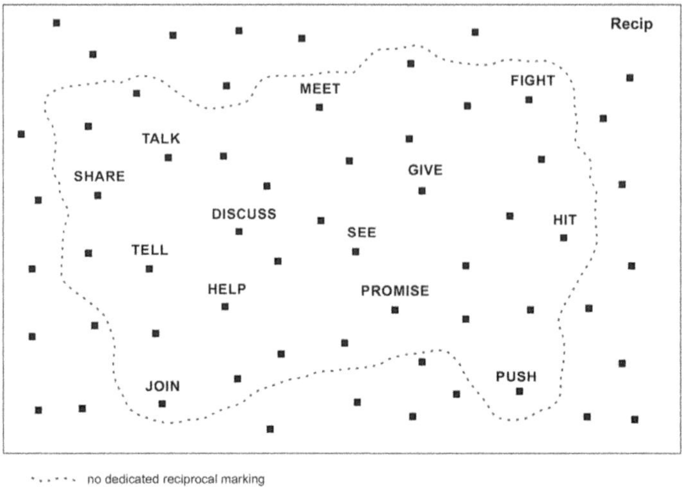

Figure 5.4 The Expression of Reciprocity in Ku Waru. The dotted line indicates that the expression of reciprocity is possible in this language, however, mostly without any dedicated formal means. Adjunct predicates (■) provide lexical compositions that may imply reciprocity given the right speech context. Ku Waru also possesses simple verbs (Rumsey, 2017); however, in the present investigation (based on the spoken text material in Merlan & Rumsey, 1991), the focus lay on adjunct predicates alone. For a full list of verbal composites with possible reciprocal meanings, see Table 11.4, Appendix B. *Source*: Author created.

Additional context factors like previous discourse, situational context, or common-ground knowledge between speaker and addressee seem sufficient to determine the meaning from one context to another.

Maybe this conclusion rests on shaky grounds. The major reasons are that we have here included a natural data set that (1) was gained from a special speech style (public oratory) and (2) is methodologically not equivalent to the more controlled elicited data sets presented for the other languages discussed earlier (Nunggubuyu, Skou, Kilivila, etc.). Moreover, we can observe that a language like English, too, shows much flexibility in the explicitness of reciprocal marking, so that quite a number of lexical items (e.g., verbs like *fight*, *talk*, *chat*, *meet*) function as "lexical reciprocals" (more on these in chapter 7) and are common, perhaps even the most common means of expressing reciprocity under natural conditions of speaking. Nevertheless, despite these reservations, we can still venture to observe that the reliance on lexical reciprocals in Ku Waru is greater than in many other languages (and formally more opaque, since Ku Waru unlike English doesn't possess a sharp distinction between transitive and intransitive sentences). Moreover, we can observe how "strategic" speakers of a language may choose to either employ or omit dedicated means in order to bring particular configurational meanings into the focus of discourse. Functional linguists and typologists may declare such observations as trivial; but it is an essential observation in order to demonstrate that such human linguistic behavior rests on an intentionality that in evolutionary terms is rooted in the tool intelligence of prelinguistic animals (see chapter 3).

Another consideration comes from linguistic-anthropological works like Rumsey (1990, 2010). These works draw attention to the fallacy of wrongly projecting structural-functional dichotomies from one language onto another. Just because some languages draw sharp formal distinctions between different meanings, other languages may not be said to "lack" them. The argument is based on linguistic relativity. The "fashions of speaking" (Whorf, 1956; cited in Rumsey, 1990) in reference to the world may simply be different to begin with. Their linguistic constructions may not imply the same kind of meaning dichotomies as they exist in other languages. At a higher level of analysis, Rumsey (1990, 2010) connects such different ways of speaking with different language ideologies that pervade languages across different functional categories (e.g., reported speech, argument structure, anaphorical/indexical reference). In the comparison of European with non-European (or non-Standard Average European) languages, it appears that literacy and higher education in particular have strong impacts on how speakers of languages perceive of standards of speaking and how some variants of speech are intuitively preferred over others in elicited speech (cf. the effects of literacy on complex sentence structure, see Mithun, 1984; a talk-action dichotomy in Standard Average European, see Rumsey, 1990; the written language bias, see Linell, 2005). From this perspective, Ku Waru might after all provide an interesting example of a language in which higher-educated speech, skillful oratory, tends away from a standard of speaking that endorses formal explicitness in the expression of reciprocity. Rather, if anything it is the other way round, namely that speakers of the language value fashions of speaking that "enact" ambiguity and vagueness.

NOTES

1. For details on notational conventions with regard to language examples, see chapter 4, endnote 1.

2. Like Nunggubuyu, the Nyulnyulan languages are non-Pama-Nyungan languages, which are traditionally spoken in the Kimberley region of Western Australia (Stokes & McGregor, 2003). As is discussed in Alpher et al. (2003, pp. 341–43), there exists a conspicuous genetic relationship between various verbal derivational suffixes in different non-Pama-Nyungan as well as Pama-Nyungan languages. In cross-linguistic perspective, however, these suffix cognates vary with regard to their functional properties: in some languages, like the Nyulnyulan languages, it carries

both reciprocal and reflexive meaning; in others, like Nunggubuyu, it only carries reciprocal meaning and in still others, it carries the reflexive meaning alone. Based on their comparative evidence, Alpher et al. (2003) argue that it is most likely that originally the proto-form of this suffix only conveyed the reciprocal function and subsequently extended its functional range in some languages to include the reflexive meaning: "The distinctive reciprocal forms in Ngn and Nu [Nunggubuyu] appear to be cognate with each other and with the combined reciprocal/reflexive forms in BGW, D, M, Ngal, and R, suggesting that, in these languages, the original reciprocal suffix has extended its range to replace the original reflexive. This is a quite plausible development cross-linguistically" (Alpher et al., 2003, p. 343).

3. Reduplication in general is a highly productive derivation process in Kâte, used to derive new lexemes as well as grammatical words. Thus, reduplication is not restricted to the formation of reciprocal word forms alone (Keyßer, 1969; Pilhofer, 1969).

4. Note that the latter two, segmentary-person reference and *ung eke* ('bent speech') are also found in oratory that is not delivered in *el ung* style but are not as frequent there (for the statistical analysis, see Merlan & Rumsey, 1991, pp. 100–2).

5. In this context, special thanks are due to Alan Rumsey who (in April 2010, in Canberra) kindly went through some of the early interlinearizations with me and thereby greatly encouraged me to proceed with them and to include them in this work. Moreover, I am indebted to him and Francesca Merlan for allowing me to include further data sets (see endnote 7) into the discussion and for continued support in my analysis of the Ku Waru language material during the realization of this book project in 2017–2018 (especially with example [16c] and the analysis of *kapu te- ti-* as a serial verb construction involving an adjunct + verb complex; see the main text). The glossing in the examples (16a–c) follows the glossing convention in Merlan and Rumsey's (1991) grammar sketch (Appendix B) except with regard to the indication of unsegmented (not further analyzed) forms. Following the Leipzig Glossing Rules, they receive glossing through a colon (Rule 4C). All remaining errors in the analysis are my own.

6. As is detailed in Rumsey (2017, pp. 102–4), Ku Waru also possesses *simple verbs* like *pu-* 'go' or *nu-* 'eat' which are frequent in adults' speech and play an important role during language acquisition in young children. In Rumsey's (2017) terminology, simple verbs are contrasted with the more complex *adjunct + verb* constructions (e.g., *nok to-* [lit. cough hit-] 'cough,' *odi le-* [lit. pour put.in.place-] 'pour'), and these in turn are contrasted with the formally (and sometimes also conceptually) more complex *serial verb constructions* (e.g., *me- o-* [lit. carry- come-] 'bring,' *kodu- nyi-* [lit. pull- say-] 'tell-').

7. The headlines of this unpublished manuscript (Merlan & Rumsey, 1989) read: "Ku Waru—English Word List; Current as of September 1989; Compiled in Lexware by Bruce Britts and Stephen Roberts from Merlan and Rumsey's card file and field notebooks." In total, this e-file document contains 2,259 word type entries. Alan Rumsey provided me with Merlan and Rumsey (1989) as an electronic text file. The permission to retrieve and publish data from this document was granted to me by Francesca Merlan and Alan Rumsey (July 2018).

8. The author owes this insight to Alan Rumsey (p.c., April 2010, Canberra).

9. For a full list of such (lexical-reciprocal) verb composites, see entries in Table 11.4 in Appendix B.

Chapter 6

Functional Variation

Prediction 2 (see chapter 3)

The logical space for reciprocal marking is greater than the empirically occupied space of reciprocal marking in individual languages. The conventionalized distribution and productivity of reciprocal markings is thus limited and language-specific. Some reciprocal markings are highly productive, extending across a wide range of reciprocal situation types. Others are less productive and are limited by constructional (syntax-related) and/or semantic and pragmatic (meaning-related) factors. Some reciprocal markings may even be entirely unproductive and limited to a very specific set of lexically defined constructions.

Prediction 4 (see chapter 3)

Reciprocal markings are likely to show multiple functions that extend beyond the reciprocal domain. Elements identified as reciprocal markers will have different historical origins and therefore possess language-specific structural and functional features. Consequently, they will differ from one another in respect to their potentials for transfer to other functional meanings that are nonreciprocal.

In the following, we discuss the different distributions and degrees of productivity of reciprocal marking patterns in individual languages. Furthermore, we consider various functional extensions to nonreciprocal meanings.

YIDINY, DYIRBAL, AND DYABUGAY

In this first part of chapter 6, we focus on three Australian Aboriginal languages (once) spoken in the coastal regions of Northeast Australia (see map

4.1, p. 68). At the time of their linguistic documentation by R. M. W. Dixon in the 1970s, they were still spoken by some tribal people living in the Cairns region (Dixon, 1972, 1977). However, the number of fluent speakers was already scarce and much of the tribal coherence of former days was disrupted. Yidiny and Dyirbal are (were) spoken in directly adjacent tribal territories south of Cairns. Their tribal neighbors to the north and south spoke Dyabugay and Wargamay, respectively (Dixon & Blake, 1991, p. 348). They all belong to the Pama-Nyungan subbranch of the Australian language family. According to Dixon (2002), Yidiny and Dyabugay together form a genetic subgroup, labeled the "Cairns subgroup," while Dyirbal and Wargamay belong to another group labeled "Herbert River group" (Dixon, 2002, pp. xxxii–xxxiii, pp. 659–60).

In Yidiny, there is only a single verb stem (tr. *bundja-* 'to hit') that can be used in a reduplicated form in a monoclausal construction in reference to a reciprocal situation (intr. *bundja-bundja-* 'to fight each other'; Dixon, 1977, p. 282). With other transitive verbs, verbal reduplication is not productive as a reciprocal marking pattern (i.e., **wiwi-wiwi-* 'give each other,' **budji-budji-* 'say to each other,' **niba-niba-* 'show (something) to each other,' **gidjar-gidjar-gunda-* 'paint each other'). The reason for this non-productivity cannot be explained in terms of any structural restriction, since verbal reduplication itself does exist in the language. However, instead of conveying reciprocal meanings, verbal reduplication with verbs other than *bundja-* 'hit' carries an association with iterativity and duration (e.g., *wawal-* 'look,' *djaŋa-* 'grumble at,' *bugan-* 'eat' or *bandjar-* 'follow,' when reduplicated, express meanings such as 'look thoroughly,' 'grumble a lot,' 'eat quickly' or 'follow quickly,' respectively). Hence, speakers of Yidiny rely on a different strategy for expressing reciprocal meanings. The standard way of doing this is to draw on bi-clausal constructions with inverse role assignment ('A Verb B, B Verb A'), involving the morphologically invariant expression (particle) *djaymbi/djaybar* 'in return; in exchange' in either clause-final or some other position within the sentence (Dixon, 1977, p. 282). This strategy is even sometimes applied to the otherwise reduplicable verb *bundja-* 'hit' (see example [17]).[1]

(17) YIDINY

ŋayu	*bama*	*bundja:ny /*	*ŋanyany*	*bama:l*	***djaymbi***
1Sg.A	person.Abs	hit.Pst	1Sg.Acc	person.Erg	in.return

bundja:ny
hit.Pst
'I hit the person and the person hit me in return.' (Dixon, 1977, p. 379)

(18) YIDINY

ŋayu *ŋunydju:nda* *guru:n* *wiwi:ny /*
1Sg.A that.Dat story.Abs give.Pst

ŋanda ***djaymbi*** *guru:n* *budji:ny* *yinydju:ŋ* *bama:l*
1Sg.Dat in.turn story.Abs tell.Pst this.Erg person.Erg
'I told that [person] a story, and then this person, in turn, told me one.'
(Dixon, 1977, p. 380)

This construction is only productive with transitive (semantically other-directed) verbs, and it usually does not involve reference to more than two (human) participants (A, B). Nevertheless, a few verbs that are semantically other-directed (e.g., *budji-* 'to speak,' *baga-* 'to spear') are not customarily used in this construction. Instead, they are simply used as bare reciprocals with only vague, implicated reference to the reciprocal situation ('they spoke,' 'they pierced'). In example (19), the a-sentence gets a transitive (other-directed) reading, while the b-sentence receives a reciprocal (mutually directed) interpretation.

(19) YIDINY

a. *ŋayu* ***budjiiny*** */ ŋudju / bama:n*
1Sg.A tell.Pst No person.Dat
'I told [him]: No, they belong to other people.' (Dixon, 1991, T1.28–29)

b. *ŋanydji* ***budjiiny*** */ ŋanydji* *garru* *warraŋadji:na*
1Pl.A tell.Pst 1Pl.A by-and-by shift camp.Purp
'We spoke [among ourselves]: We must shift camp by-and-by.'
(Dixon, 1991, T5.4)

Finally, Yidiny exhibits a very special type of construction: the use of the nominal suffix *-bany* with spatial expressions ("locational qualifiers"), but only in reference to dyadic, symmetrical, non-dynamic relations between human participants (e.g., *nyarru:-bany* '[two people sleep] on top of each other'; *walmbi:rr-bany* '[two people sit] next to each other, side-by-side'; *garba-garba:-bany* '[two people stand] behind each other, back-to-back'). "The affix *-bany* indicates that two individuals are in juxtaposition; the root to which *-bany* is affixed marks the positional relation between them" (Dixon, 1977, p. 161). Only in a very few cases is this nominal suffix also attested with nonhuman referents (e.g., a pair of stirrups hanging alongside each other) and with other nominal expressions, which have assumed idiomatic meanings in the domain of social kinship relations (*djili:-bany* 'adult and closely related child') and social exchange (*gulgagulga:-bany* 'to share out, divide'). However, Yidiny *-bany* is never a reciprocal marker on verbs.

We do not need to look far for languages that set very different standards of reciprocal marking. Let us consider the two neighboring languages of Yidiny: Dyirbal and Dyabugay. In Dyirbal, verbal reduplication plays a much more prominent role in reciprocal marking than it does (did) in Yidiny. It is obligatorily and productively used in this function with transitive verb forms, and it is always combined with the verbal suffix *-(n)barriy* (Dixon, 1972, p. 92). Resulting formations are, for instance, *djurrgay-djurrgay-barriy* 'spear each other' (see example [20]), *nyundjal-nyundjal-nbarriy* 'kiss each other,' *djaymbal-djaymbal-barriy* 'find each other; meet.'

(20) DYIRBAL

balagarra	*bayi*	*yara*	*djurrgay-djurrgay-**barrinyu***
two people.Nom	there.Nom.I	man.Abs	Rdp-spear-Recip.Pres/Pst

'The two men are spearing each other.' (Dixon, 1972, p. 92)

Thus, in contexts where Yidiny speakers are relying on the bi-clausal construction type ('A Verb B, B Verb A'), Dyirbal speakers are employing a morphologically dedicated, monoclausal construction. The bi-clausal construction also exists in Dyirbal, but speakers do not customarily draw on it when expressing reciprocal meanings as outlined above. Rather, Dixon clarifies, the reciprocal particle *ŋurri* "indicates that an event happens to redress a balance, for example as revenge; or as a gift from A to B in return for an earlier gift from B to A; or someone taking his turn at doing something" (Dixon, 1972, p. 120).

Yet, even the reduplicating verbal marking pattern ('Rdp-V + *-(n)barriy*') is not applicable to all verb forms in Dyirbal. One of them is the verb *bundalay* 'fight,' which Dixon identifies as "probably [being] a fossilised reciprocal form" (Dixon, 1972, p. 299). Interestingly, this form seems etymologically related to Yidiny *bundja-* 'hit,' which is the only verb form in that language which can be used reciprocally by way of verbal reduplication (see above). The other example of a "fossilised reciprocal form," or "lexical reciprocal" (Nedjalkov, 2007b; Knjazev, 2007), is the verb *munydjal* 'divide,' which behaves intransitively and always requires a plural subject. Besides this intransitive verb form, there are eight other verbs of giving (e.g., *wugal* 'give'), all of which are transitive and require explicit reciprocal marking (Dixon, 1972, p. 300). It is again interesting that Dyirbal should have a lexical reciprocal in a semantic domain where Yidiny, too, makes use of a special word form, namely the complex expression *gulgagulga:bany* 'share out, a piece to every one.' However, again, the encoding of mutuality has not been derived by the same formal means: in Dyirbal, the verb form alone has come to denote a mutual sharing event, while in Yidiny a complex verb form has been derived and lexicalized by means of nominal reduplication and suffixation with nominal *-bany* (see above).

In Dyabugay, finally, verbal reduplication does not play a role in reciprocal marking at all. Here, the verbal suffix *-lndjirri/-nydjirri* alone carries that function (Patz, 1991, p. 283, pp. 296–97). The reciprocal suffix is mainly productive with transitive verb roots like *ngunda-l* 'watch, look at,' *buga-y*

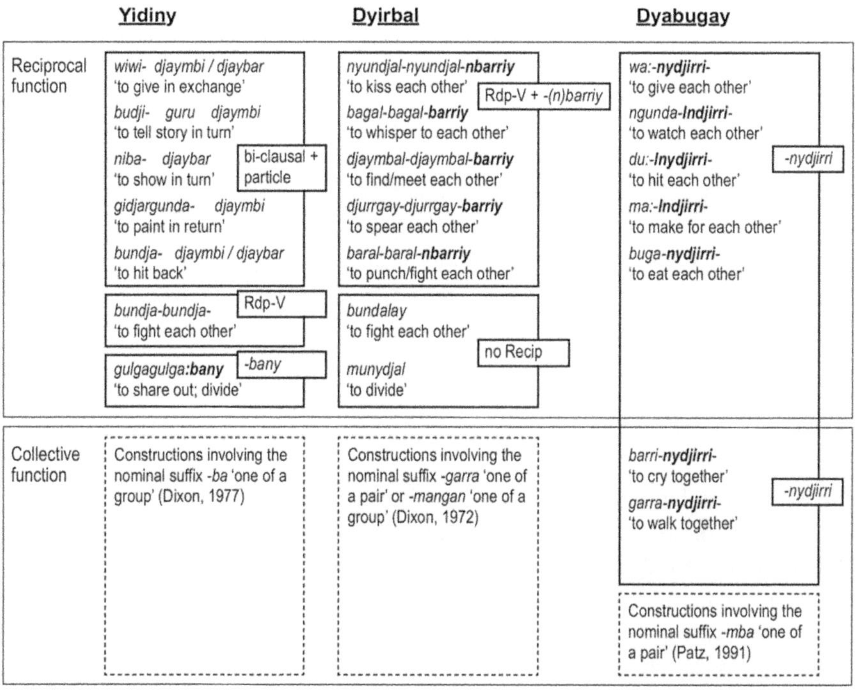

Figure 6.1 The Form and Distribution of Reciprocals in Three Neighboring Australian Languages. Yidiny has no productive and dedicated reciprocal marking pattern other than the bi-clausal construction type ('A Verb B, B Verb A in return'). Verbal reduplication as a means of expressing reciprocal meanings is only applicable to a single verb stem (*bundja-* 'hit'). Dyirbal, on the other hand, possesses a fully productive and formally dedicated reciprocal marking strategy involving verbal reduplication and suffixation with verbal *-(n)barriy*. However, this marking strategy is not productive with (inherently reciprocal) verbs like *bundalay* 'fight' and *munydjal* 'divide.' In Dyabugay, in turn, verbal reduplication plays no role at all in the expression of reciprocal meanings; moreover, the verbal suffix marker *-nydjirri* is not only productive with transitive verbs as a marker of reciprocal action but also with intransitive verbs denoting collective action. For references of all figure entries see Table 12.2 in Appendix C. *Source*: Author created.

'eat,' *du:-l* 'hit, kill' and *wa:-y* 'give.' However, in addition and therefore contrast to Dyirbal, it is also productive with intransitive verbs like *barri-y* 'cry' and *garra-y* 'come,' where it marks collective ('acting together') rather than reciprocal ('acting upon each other') action. In Dyirbal, the verbal suffix *-(n)barriy* is only applicable to transitive verb stems. In Dyabugay, the suffix is productive with both transitive and intransitive verb stems. Due to this latter extension, the Dyabugay suffix also allows readings outside the standard domain of reciprocity. Possible conceptual motivations for such functional extensions are discussed in chapter 8 (section on "Reciprocal Schemas"). For now, we merely summarize the observed cross-linguistic differences in the structural and functional encoding of reciprocity in Figure 6.1.

The data show that even closely related languages spoken in neighboring territories (Yidiny, Dyirbal, Dyabugay) do not converge on identical applications in the use of reciprocal markers. Speakers of each language follow different customs of conveying reciprocal meanings to one another. This difference in custom extends equally across form and function of language use. In other words, the difference does not just concern the structural features, it also concerns their functional ranges of application (e.g., dynamic versus static events; reciprocal versus reciprocal-collective events) as well as their productivity with specific lexical items. Figure 6.1 summarizes these facts schematically.

THE MANTHARTA AND NGAYARDA LANGUAGES

The Mantharta language, consisting of four dialects (Tharrkari, Warriyangka, Jiwarli, Thiin), and the Ngayarda languages, to which (among others) belong Martuthunira, Yindjibarndi, and Panyjima, are spoken in the Pilbara region, Northwest Australia (see Map 4.1, p. 68). They are all classified as Pama-Nyungan languages. The Ngayarda languages form a genetic subgroup (Dench, 1987), while the genealogy between the Ngayarda and Mantharta languages is less clear. Dixon (2002) subsumes them, together with the Kanyara languages, under the label "Gascoyne River to Pilbara areal group" (p. xxxviii) and explains with regard to the term *areal group*: "Some of what have been suggested as genetic subgroups are in fact small linguistic areas whose member languages appear not to be closely genetically related but to have been in contact for a considerable period and as a result a number of linguistic traits have diffused over the area" (p. 659). The tribal territories belonging to Martuthunira, Yindjibarndi, and Panyjima are more or less adjacent to one another; the Mantharta dialect Jiwarli is spoken in territories a bit further to the south.

As is pointed out by Dench (1987), "[t]here is a remarkable degree of cognacy among reflexive/reciprocal forms across the Australian continent. For example, notice the similarity among the Ngayarda forms *-yarri*, *-lwarri*, and *-nmarri*, and the Dyirbal forms *-yirri* (reflexive) and *-nbarri* (reciprocal)" (p. 338). Similarly, Dixon (2002) discusses the reconstructed reflex form **-dharri*[2] and its pervasiveness in modern Australian Aboriginal languages:

> There is a recurrent derivational suffix, which probably had [the] original form **-dharri* and has become, by assimilation and reduction, *-dhirri*, *-dhi*, *-dji*, *-yi*, *-rri*, etc., in modern languages. This is found in languages right around the continent—as a derivational suffix—which suggests that is has a long history as a suffix, in contrast to the various causatives and applicatives, which probably evolved (from verbs) as suffixes fairly recently. [new paragraph] Reflexes of **-dharri* have varying functions but generally include reflexive and sometimes also reciprocal. (Dixon, 2002, pp. 206–207)

As was mentioned in chapter 5 (section "Nunggubuyu and Other Languages" including endnote 1), Alpher et al. (2003) likewise offer a diachronic analysis of the reflexive/reciprocal verbal suffixes found in many non-Pama-Nyungan as well as Pama-Nyungan languages. Just like in Dixon (2002), their focus is on verbal suffixes with either reflexive or reciprocal function. Moreover, they focus on a genetic subgroup in Northern Australia (Gunwinyguan).

In the Mantharta and Ngayarda languages, however, the situation is strikingly different. In most (if not all) languages the verbal suffix is referred to as the *collective suffix* (glossed as "Coll" instead of "Recip"), since apart from marking reciprocal actions it also expresses *collective* actions ('A and B performing actions together'). In all four languages, the collective suffix has different allomorphic realizations, depending on the conjugation class membership (and length) of the verb stem. With the exception of Panyjima, all of these forms end in /-rri/. Moreover, in three of the four languages (again, with the exception of Panyjima), the collective form *-marri* can be used on nominal roots in order to derive collective-reciprocal verb forms. Alan Dench, a long-term expert on this particular linguistic region, cautions us with respect to even the closely related Ngayarda languages: "While the suffixes in the different languages appear to be cognate, it is not clear that a single set of forms should be reconstructed for proto-Ngayarda" (Dench, 1987, p. 323). Despite such difficulties in the diachronic analysis, we now turn to each language and venture a systematic cross-linguistic comparison from the synchronic point of view.

In Jiwarli, the collective suffix forms *-lparri*, *-tharri*, *-alyurri*, *-marri*, and *-rri* are highly productive with both transitive and intransitive verbs (Austin,

1992, 1997). Transitive verb stems marked by *-( )rri* and used in construction with a plural subject are typically interpreted as reciprocal in meaning ('people acting upon each other'). Intransitive verb stems marked by the same suffix and used in construction with plural subjects are typically interpreted as collective in meaning ('people doing something together'). Here, the collective suffix may even take inanimate subjects as in the use of *kampa-alyurri-* '(fires) burning together' (Austin, 1997, T11.1, elicited material by P. Austin, p.c.). However, the suffix does not always derive *reciprocal* verb forms from transitive stems. In combination with verbal complements like a dative object or a comitative NP, the derived forms—which would otherwise (with only plural subject) be interpreted as reciprocal (e.g., 'they fought-Coll against each other,' 'they laughed at-Coll each other')—may assume a collective or comitative sense (i.e., 'they fought-Coll against the enemy' or 'they laughed-Coll with their daughters') (see examples in Austin, 1997, T46.5, T48.33). The Jiwarli collective suffix may therefore also occur in quasi-transitive constructions with a collective (rather than reciprocal) interpretation.[3]

In Martuthunira, the collective suffix forms *-lwarri*, *-yarri*, and *-marri* are also fully productive with both transitive and intransitive verb stems. Despite the general tendency to use the collective suffix forms as a marker of reciprocity with transitive stems, and a marker of collective action with intransitive stems, Dench (1995) maintains that Martuthunira shows no restrictions against the collective reading with transitive stems: "There is no syntactic context which forces a reciprocal reading for a verb bearing the collective suffix. Although a reciprocal reading is available where a transitive verb appears with no object and the subject is non-singular, a collective reading is always possible here given the frequent ellipsis of arguments" (Dench, 1995, p. 155). Accordingly, transitive verbs marked by the collective suffix may either carry a reciprocal or a collective sense ('people acting towards/upon each other' versus 'people acting together towards something/someone').

Moreover, Martuthunira differs from the other languages by its highly prolific application of the collective suffix in both plural and singular subject constructions with no, from the European perspective, obvious either collective or reciprocal interpretation. Thus, it also occurs in sentences with meanings like 'the snakewood in this country is-Coll useless' (T7.264), '[they] sent-Coll word to those people' (p. 220), 'that man didn't come to bury-Coll his grandson in the ground' (p. 107), or 'How about you give-Coll one [of your cakes] to this fellow?' (T4.54) (see Dench, 1995). Dench (1987) dedicates an entire paper to the discussion of such occurrences of the collective suffix, arguing that in Martuthunira (and Panyjima as well) it has assumed the additional function of a special *kinship marker*. The participants referred to denote people belonging to the same set of merged alternate generations (e.g., a man and his grandson, a man and his grandfather, a man and his brother-in-law). According to Dench, the collective suffix can only occur in such cases where the persons referred to stand in this kind of kinship relation to one another; it cannot be used when the participants belong to different generation sets (e.g., parent and child, parent and parent's father, man and mother-in-law). Cross-linguistically, this phenomenon seems very rare.

In Yindjibarndi, the collective suffix forms *-nmarri*, *-rnmarri-*, *-marri*, and *-nytyarri* are, again, fully productive with transitive and intransitive stems, deriving collective-reciprocal verb forms with plural subjects (Wordick, 1982). Unlike the Jiwarli and Martuthunira suffixes, the Yindjibarndi suffix shows a clearer preference for either collective or reciprocal readings depending on the valence of the underlying verb stem: with intransitive verb stems, the derived meaning is collective ('together'); with transitive stems, the interpretation of the derived verb form is reciprocal ('each other'). With transitive verb stems at least, the Yindjibarndi collective suffix has a clearly detransitivizing effect, which disprefers the collective interpretation. Wordick (1982) does not explicitly mention an extended use of the collective suffix in the domain of transitive singular subject constructions in the context of kinship marking (though Dench, 1987, considers the possibility that such a usage may once have existed in the language).

In Panyjima, the collective suffix appears in two distinct forms and functions, which are labelled by Dench (1991) as *Collective* and *Collective2*, respectively. The closest structural and functional equivalent to the other Ngayarda and Mantharta languages is the *Collective* suffix, with allomorphic variants *-nyayi/-(nma)yi*. It is productive with both transitive and intransitive verb stems and expresses both reciprocal and collective meanings. However, in contrast to the other three languages, it is most productive as a marker of reciprocity with transitive verb stems, while only rarely used as a marker of collective action with intransitive stems. Indeed, by far the majority of the derived verb forms found in Dench (1991) are reciprocal in meaning and derived from transitive stems. For this reason, Dench refers to the verbal suffix as reciprocal (rather than collective) marker (Dench, 1981, p. 120; Dench, 1987, p. 338). He thereby foregrounds its current synchronic use and function, rather than its shared historical identity with the cognate markers of the neighboring languages.

In addition, just as in Martuthunira, the *-nyayi/-(nma)yi* forms in Panyjima also occur in transitive singular subject constructions to mark special forms of social kinship (e.g., 'Did you see-Coll your brothers?'). In this function, the suffix neither influences the transitive status of the verb stem, nor does it affect the other-directed meaning of the sentence ('Did A see-Coll B?').

The other collective marker in Panyjima, *Collective2*, is the invariant form *-rri*. Phonologically, this verbal suffix is very similar to the collective suffixes in the other three languages. However, its productivity and interpretation are much more limited and functionally biased (Dench, 1991, p. 192). It is only productive with a very few verb stems, and invariably forces a collective 'together'-meaning onto a given verb stem, even with transitive stems (e.g., *yinya-rri-* give-Coll 'give together (to s.o.)'; *thitha-rri-* wait.for-Coll 'wait together for'; *thurni-rri-* laugh.at-Coll 'laugh together at'). This restricted use stands in stark contrast to the prolific uses of the collective suffixes in Jiwarli and Martuthunira, where both collective and reciprocal readings are (potentially) available with transitive verb stems. And where, moreover, other factors like the specific verb stem, the argument structure of the sentence, or the surrounding discourse, also play a key role in determining the meaning of the derived form.

In Figure 6.2, we capture schematically some of the similarities and differences in distributional pattern as they have been outlined for the collective suffix in the previous passages. As can be seen, most of the collective suffixes—excepting the form *-rri* in Panyjima—extend their use across at least two different functional domains (mostly Recip and Coll) and are thus not restricted to only a single grammatical function. However, as is outlined above, they differ from one another in the extent to which they are productive as markers of such different functions. For instance, the use of *-marri* as marker of collective action is highly productive in both Jiwarli and Martuthunira, where it is also available in this function with transitive (not just intransitive) verb stems. By contrast, the collective use of *-marri* in Yindjibarndi and that of *-nyayi* in Panyjima is much more limited in this respect, allowing a collective sense only with intransitive verb stems, and otherwise only allowing a reciprocal interpretation with most verb stems (this is at least partly indicated in the figure by a more limited frame-extension into the "Coll" domain when compared with the other two languages).

Another notable similarity/difference (indicated in Figure 6.2) is the differential extension of the collective suffix into the domain of social kinship marking in singular-subject constructions in both Martuthunira (*-marri*) and Panyjima (*-nyayi*). This does not apply to the other suffixes in both Jiwarli and Yindjibarndi, where the occurrence of *-marri* remains restricted to plural subject constructions. Finally, the schematic representation illustrates how Panyjima alone employs two different verbal suffix forms for the expression of reciprocal/collective actions: the form /-rri/, which is *phonologically* most similar to the other collective suffix forms (/-marri/) represented to the left, is only productive as marker of collective action, even with transitive verb stems; the form /-nyayi/, on the other hand, which is *phonologically* more

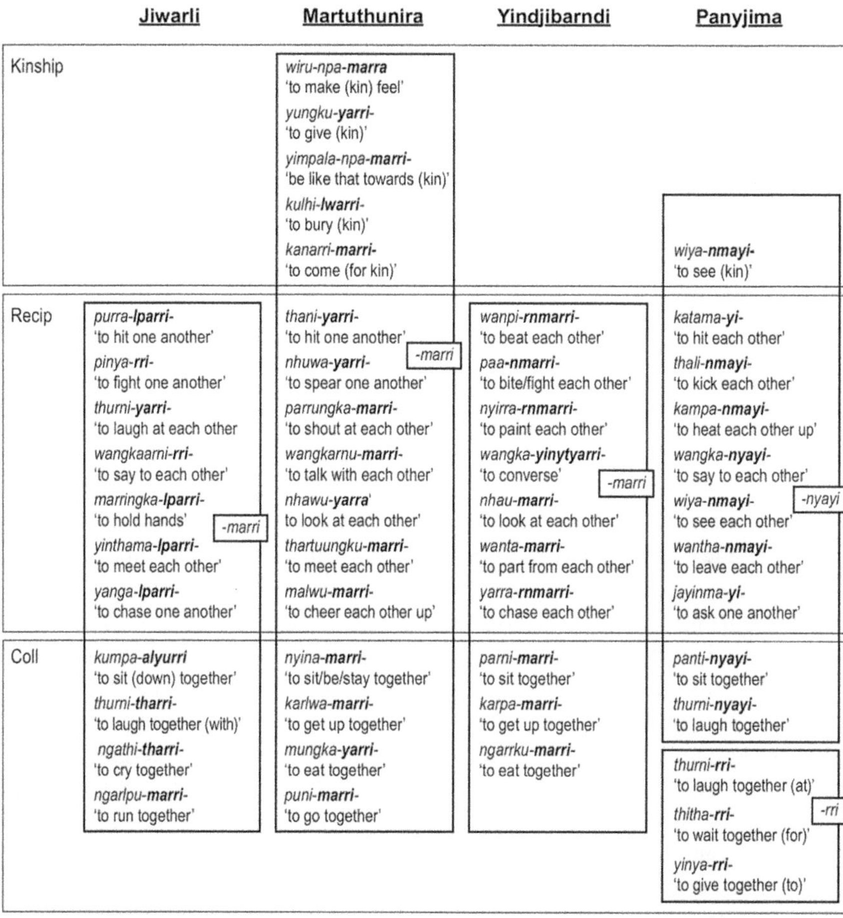

Figure 6.2 Different Functional Ranges of Verbal -( )*rri*. Schematic representation of the different distributional patterns displayed by the collective suffix in different members of the Mantharta and Ngayarda language group. Each black frame (labeled "Kinship," "Recip," "Coll") represents a specific function (i.e., configurational meaning) for which there are (might be) dedicated linguistic expressions in a language. The reciprocal function, as the focus of our attention, is represented in the mid-section; the collective function is represented in the frame below and the collective kinship function as identified by Dench (1987) is represented in the topmost frame. The vertical frames in turn represent the different patterns of distribution—across these different functional domains—as they are displayed by the collective suffix in each individual language. For detailed references of all figure entries see Table 12.3 in Appendix C. *Source*: Author created.

distinct from the other collective suffix forms, is primarily productive as marker of reciprocal action. In its overall *functional and structural* distribution, it therefore more closely resembles the distributional patterns of *-marri* in the neighboring languages.

NUNGGUBUYU AND NYULNYULAN LANGUAGES

We return to Nunggubuyu, the non-Pama-Nyungan language in Northern Australia already discussed in chapter 5 (see Map 4.1, p. 68), this time with the focus on functional variation. The verbal suffix *-nyji* in Nunggubuyu can be used in a number of different functions—although with various degrees of productivity. Among the different functions are the reciprocal, the reflexive, and the collective meaning. Since the suffix is by far most productive as a marker of reciprocity, it is habitually referred to as the reciprocal suffix of the language. In other words, nonreciprocal meanings like the reflexive and the collective meaning are mostly expressed by other formal means than the reciprocal suffix. There are only a few instances where *-nyji* is used to derive such nonreciprocal meanings. The lists of examples which can be given for nonreciprocal uses of *-nyji* are therefore exhaustive.

For example, as a marker of reflexive, self-directed action (e.g., 'he hit himself'), *-nyji* only occurs in construction with three transitive verb roots (Heath, 1984, p. 391): *=wa-* 'to bite' (Refl *=wa-ngi-nyji-* 'to bite oneself'), *=r̲a-* 'to spear' (Refl *=r̲i-nyji-* 'to spear oneself') and *=yami-* 'to say this' (Refl *=yami-jga-nyji-* 'to say this to oneself'). Interestingly, in each case, the verb marked by *-nyji* is also productive as a reciprocal verb form. The preferred reading mainly depends on the number of participants encoded by the subject: with a singular subject, only the reflexive reading is possible (one person acting upon themself); with a plural subject, the reciprocal reading tends to be preferred (several persons acting among each other), though the specific context may also favor a reflexive interpretation (several persons acting each unto themself).

Similarly, *-nyji* as a marker of collective ('together') action, or else by Heath referred to as *joint involvement*, is only found in construction with three intransitive verb roots (Heath, 1984, p. 392): *=lhagarara-* 'to make loud noises' (Coll *=lhagar̲ari-nyji-* '(group) to make loud noises together'), *=lida-* '(dog) to play around' (Coll *=l̲idi-nyji-* '[dogs] to frolic together') and *=milha-* 'to shine' (Coll *=milhi-nyji-* 'to shine together').

As a marker of reciprocal meanings, on the other hand, *-nyji* is highly productive. Heath observes: "The Recip formation is rather productive; there are many verbs which do not have it but this seems to be because it would not make sense logically with them (rather than because of arbitrary lexical restrictions)" (Heath, 1984, p. 393). Since the suffix is so highly productive in the language, it must be assumed that there exist many combinations of 'transitive verb stem plus *-nyji*' yielding a reciprocal interpretation, too many to be listed in the dictionary entries of the language—notwithstanding Heath's (1981) very comprehensive documentation of lexical items and derived formations (in my count yielding 71 reciprocal *-nyji*-formations; see entries in Tables 11.1.1–11.1.6, Appendix B).

While in Nunggubuyu the verbal suffix *-nyji* is thus only truly productive as a marker of reciprocity, we find that in Nyulnyulan languages again the situation is quite different. Recall from chapter 5 that this small family of non-Pama-Nyungan languages in Northwestern Australia (see Map 4.1, p. 68) possesses a partly cognate but morphologically more complex marking pattern involving the cognate verbal suffix marker *-nyji/-inyj* in combination with the prefixal element *ma-, mi-,* or *m-* (McGregor, 1999). In this language group, the marking pattern is not only productive as a marker of reciprocity, it is also fully productive as a marker of reflexivity. Below are given some examples from Warrwa (see [21] and [22]) with *ma-V-nyji* as a marker of either reflexive or reciprocal action.

(21) WARRWA

*Ngayu nga-**ma**-ng-ka-**nyji**-ny.*
1:Min 1:Nom-RR-E:Nasal-hit-RR-Pst
'I hit myself.' (McGregor, 1999, p. 89)

(22) WARRWA

*Ngi-rr-**ma**-ng-ka-**nyji**-na.*
3:Nom-Aug-RR-E:Nasal-hit-RR-Pst
'They were fighting with each other.' (reciprocal)
'They hit themselves.' (reflexive) (McGregor, 1999, p. 89)

Just as in Nunggubuyu (see above), the interpretation of a given verb form varies with the number of participants encoded in subject position. With singular subjects, only the reflexive reading is conceivable. With plural subjects, on the other hand, either the reflexive or reciprocal reading are possible. Interestingly, however, the situation is still more language-specific. For example, in sentence (23), only the reciprocal interpretation is possible, even though on principle the reflexive reading is also conceivable ('they two argued with themselves'). However, McGregor (1999, p. 90) points out that in Warrwa this interpretation is not available, not even with singular subjects (*'he argued with himself').

(23) WARRWA

*Ngi-rr-**ma**-yama-**nyji**-ny-bili.*
3:Nom-Aug-RR-argue-RR-Pst-Du
'They two argued with each other.' (McGregor, 1999, p. 89)

In the case of the Nyulnyulan marking strategy (*ma-V-nyji*), it is thus not possible to reserve the use of verbal *-nyji* (or *-inyj*) for the reciprocal function alone: it is as much a reflexive marker as it is a marker of reciprocity.

A striking contrast to the situation previously outlined for the remote neighbor to the north, Nunggubuyu. As was outlined above, the only cases where *-nyji* expresses reflexive meanings is in construction with a handful of transitive verbs, for which, Heath (1984) reasons, the use of the actual, productive reflexive marker is morphologically (or lexically) blocked: "In general, mono-syllabic verb roots either do not have or make little use of Refl derivations" (Heath, 1984, p. 391). Otherwise, reflexive meanings are productively

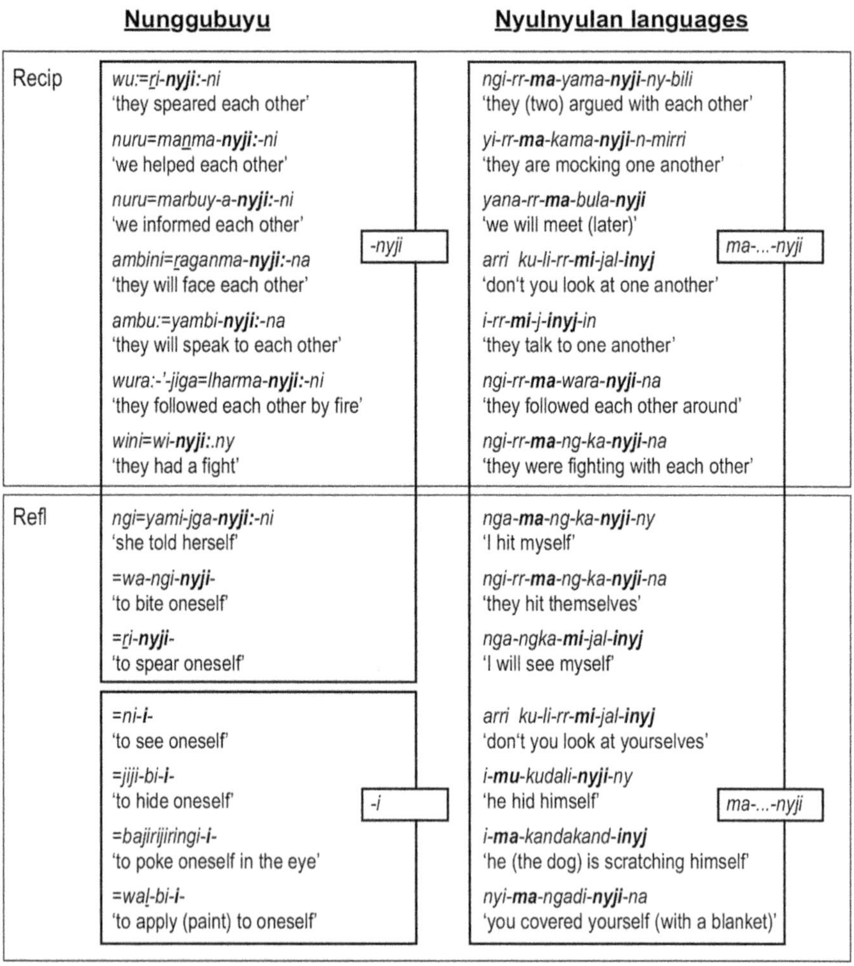

Figure 6.3 Different Functional Ranges of Verbal *-nyji*. Schematic representation of the different ranges of productivity of verbal *-nyji* in Nunggubuyu (Heath, 1984) and Nyulnyulan languages (McGregor, 1999). For detailed references of all figure entries see Table 12.4 in Appendix C. *Source*: Author created.

expressed by the verbal suffix marker *-i/-'-* (e.g., *=jiji-bi-'-* 'to hide self,' *=balhi-'-* 'to cut self,' or *=ni-'-* 'to see self').

Figure 6.3 summarizes these comparative observations: both Nunggubuyu and members of the Nyulnyulan language family (e.g., Warrwa, Nyikina, Yuwuru, Bardi) employ extant variants of the verbal suffix marker *-nyji* (in the case of Nyulnyulan languages, this suffix marker also typically co-occurs with variants of the verbal prefix *ma-*). As shown by the examples within the vertical frames, both these reciprocal markers are systematically employed as markers of reciprocal action. Both Nunggubuyu *-nyji* and Nyulnyulan *ma-V-nyji* occur productively as reciprocal markers with a wide array of transitive verb stems denoting different event types, like antagonistic action ('fighting each other'), joint action ('speaking to each other'), or physical action ('following each other').

However, as indicated by the vertical frame extensions on either side, in Nunggubuyu and the Nyulnyulan languages, the range of productivity of both these markers does not strictly coincide with the boundaries between separate functional categories like reciprocal (Recip) and reflexive (Refl) function. Instead, on both sides, verbal *-nyji* can also be seen to function as a marker of reflexive (self-directed) actions in construction with singular (or plural) subjects and transitive verb stems. Thus, on both sides, use of the verbal *-nyji* suffix is not strictly determined by one functional category alone (reciprocal function) but extends *across* the boundary and into another functional domain (reflexive function).

And finally, despite these similarities in functional extension, Nunggubuyu and the Nyulnyulan languages still do not show the same pattern of distribution: in Nunggubuyu, verbal *-nyji* is only marginally productive as a marker of reflexive action, namely only in those cases where the standard reflexive suffix *-i* is not applicable to certain transitive verb stems. In the Nyulnyulan languages, on the other hand, the verbal suffix marker is highly productive as a marker of reflexive actions; to such an extent that it is considered the standard way of expressing reflexive meanings in these languages. This means that in terms of productivity and range of application, the reflexive function is as much associated with the verbal *ma-V-nyji* construction in Nyulnyulan languages as it is with the reciprocal function. Accordingly, the vertical frame indicating the range of productivity of *ma-V-nyji* in Nyulnyulan languages is equally distributed across both functional domains, Recip and Refl. This contrasts sharply with the range of productivity indicated for the formal equivalent *-nyji* in Nunggubuyu: here, the vertical frame only marginally extends into the domain of reflexivity and is instead mostly replaced by the more standard reflexive form *-i* (see vertical frame in the horizontal Refl frame).

Heath (1984) does not offer an explanation as to the historical origin of reflexive *-i* in Nunggubuyu. However, given the fact that in the distantly related Nyulnyulan languages the marking of reflexive and reciprocal meanings is subsumed under one and the same form, namely *ma-V-nyji*, it is tempting to speculate that the reflexive marker *-i* in Nunggubuyu is descended from a historical form **-nyji*, which at some point in the history of Nunggubuyu functioned as both reflexive and reciprocal marker, just as it still does today in the Nyulnyulan languages. In this view, the historical reflex **-nyji* of Nunggubuyu would at some point have developed into two formally distinct forms: one keeping the form /-nyji/ in the context of reciprocal marking, and one eroding into the phonologically more reduced form /-i/ for the expression of reflexive meanings. This historical development would have been specific only to the history of Nunggubuyu, and perhaps some of its neighboring languages, but not to that of the more distantly related Nyulnyulan languages.

However, the situation seems more complex than that. I must acknowledge my ignorance in this specific field of historical reconstruction. Based on a much more thorough and reliable comparative analysis, Alpher et al. (2003) offer a quite different interpretation. With a focus on Gunwinyguan languages, to which Nunggubuyu belongs, they reconstruct "a very old contrast between reflexive **-yi-* and reciprocal **-nyci-*" (Alpher et al., 2003, p. 342), which leads them to conclude that the distinctive reciprocal function of *-nyji* in Nunggubuyu comes functionally closest to the original proto-form. They consequently argue that non-Pama-Nyungan as well as Pama-Nyungan languages with combined reciprocal/reflexive forms of **-nyci-* represent instances where "the original reciprocal suffix has extended its range to replace the original reflexive" (Alpher et al., 2003, p. 343). In other words, languages such as the Nyulnyulan languages with a combined reflexive/reciprocal marking pattern (*ma-V-nyji*) do not represent the original state of a reconstructed ancestor (proto-language); they represent an innovated state where a former reflexive form (**-yi-*) has gone out of use as a reflexive marker and been replaced by an extended use of the reciprocal marking pattern (**-nyci-*). Alpher et al. (2003) consider this as "a quite plausible development cross-linguistically" (p. 343).[4]

It is not within the scope of the present study to resolve the issues of this diachronic analysis. For our own purposes, it suffices to show that many Australian linguists agree that linguistic forms, even within the domain of core grammatical features, are historically related in different language groups and yet exhibit very different functional properties under synchronic conditions of usage. We have demonstrated this here extensively with regard to the various synchronic variants of the derivational verbal suffix **-dharri* (Dixon, 2002) in different language groups across the Australian continent. The range of functions associated with this form across different languages is impressive. It is even much more extensive than could be demonstrated here due to the special focus on the reciprocal function. Only in those instances where the reciprocal function was involved did we attend to synchronic variants of this form. However, as the literature makes clear, the actual range of functions is in fact much broader across languages (see Dixon, 2002, pp. 206–7, pp. 531–36; Alpher et al., 2003, pp. 341–43). In other words, in some languages, it is not even associated with the reciprocal function at all. This finding lends plausibility to the tool hypothesis because it highlights the immense flexibility in the association between form and meaning in language use. Such *means–ends dissociation* is one of the central defining features of behaviors based on *tool intelligence* (see chapter 3) and natural languages provide ample evidence for this phenomenon even in the domain of grammatical marking.

NOTES

1. For details on notational conventions with regard to language examples, see chapter 4, endnote 1.

2. Note that the asterisk (*) preceding the suffix form *-dharri* indicates a reconstructed form, which is unattested but through reconstruction of a proto-language (the common ancestor of all the languages in a given family) assumed to have once existed. It thus indicates a hypothetical form but not as in other examples an ungrammatical form.

3. Unlike the Ngayarda languages (Martuthunira, Yindjibarndi, and Panyjima; see below in the main text), Jiwarli retains a formal distinction between normal transitive constructions and derived reciprocal-collective constructions with animate nouns (Austin, 1997). In normal transitive use, the subject of a transitive verb is marked by ergative case (e.g., 'child-Erg plays with this knife'; Austin, 1997, T7.1), whereas in collective-reciprocal use, the subject of the derived verb form is marked as intransitive by absolutive case, which in Jiwarli receives zero marking (i.e., 'children-Abs are playing in the water'; elicited material by P. Austin, p.c.).

4. It is not certain, though, that this is an adequate assessment with regard to the cross-linguistic observations at large. Several other typological studies have argued that the opposite direction of functional extension, from reflexive to reciprocal is much more common cross-linguistically (Kemmer, 1993; Heine & Miyashita, 2008; Maslova, 2008). However, this is not really important here (besides, Alpher et al.'s formulation "quite plausible" simply implies that it is a possible development, not necessarily the most pervasive one in cross-linguistic perspective). What is important is the fact that functional extensions from one (a-priorily defined) grammatical category to another occur at all. In fact, they are pervasive in the development of natural languages and therefore constitutive of linguistic behavior. It is this aspect, this empirical finding, that allows us to conclude that natural languages are not based on pre-established categories (Haspelmath, 2007a) and that instead they are based on a general tool intelligence that creates structural and functional categories in socio-cultural and historical settings (see chapter 3).

Chapter 7

Family Resemblances

From the cross-linguistic perspective, it is not possible to reduce reciprocal markers to a distinct structural category such as "verbal affix," "pronoun," "particle," or "auxiliary verb"; nor to a distinct morphological process such as "verbal or pronominal reduplication"; nor to a distinct combination of such structural elements with one another. Some languages even lack dedicated reciprocal marking. Any attempt to set up a cross-linguistically valid category fails in the face of evidence from one or the other language. In other words, languages differ with respect to the structural characteristics of their reciprocal marking patterns to such an extent that it is impossible to summarize them under a single (universal) structural rule. "Theorists often resist it, but the cross-linguistic evidence is not converging on a smallish set of possibly innate categories. On the contrary, almost every newly described language presents us with some 'crazy' new category that hardly fits existing taxonomies" (Haspelmath, 2007a, p. 119).

The aforementioned citation from the German typologist Martin Haspelmath shows that this finding is not new. Concerted efforts of field linguists working on individual languages around the world, and typologists developing general paradigms for the analysis and comparison of such data sets, have shown over the past decades that such cross-linguistic structural variation exists for virtually every major grammatical category (e.g., tense, aspect, mood, causativity, possession, reflexivity, reciprocity). This finding also applies to the functional domain and the range of pragmatic-semantic factors influencing the productivity and distribution of specific marking patterns in individual languages. Hence, as was illustrated in chapters 5 and 6, languages possess not only different structural means of encoding reciprocity; their distributional patterns and degrees of productivity also vary and, in many ways, remain language-specific within the functional domain (see Figure 7.1).

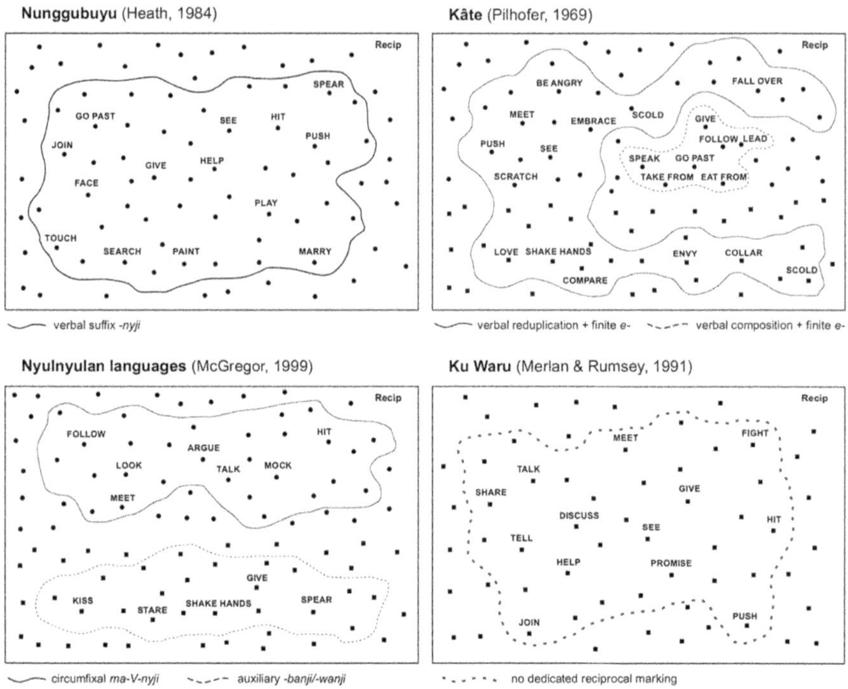

Figure 7.1 Linguistic Diversity. Reciprocal marking patterns in four different languages or language groups (Nunggubuyu, Kâte, Nyulnyulan languages, Ku Waru; see Figures 5.1–5.4 in chapter 5). Languages vary in form, productivity, and functional ranges with regard to the expression of reciprocity. Some, like the Papuan language Ku Waru, even lack dedicated formal means (see qualification of this claim in the discussion toward the end of chapter 5). *Source:* Author created.

Nevertheless, it is also true that one often converges on strikingly similar observations. Here are some examples from the marking of reciprocity:

- Languages often possess a dedicated reciprocal construction in reference to situations involving a plurality of human participants engaged in dynamic social interactions with one another (e.g., mutual hitting, fighting, helping, seeing, talking, meeting, giving);
- Languages often possess a group of lexical items (verbs) which may be used reciprocally without explicit reciprocal marking (e.g., as in English *They kissed* meaning ‘they kissed each other’ or *They met* meaning ‘they met each other’);
- Reciprocal marking patterns often extend beyond the domain of reciprocal marking and cover meanings in other functional domains (e.g., the expression of reflexive, self-directed action, or else joint, collective action);

- Reciprocal markers often have similar structural limitations, for example, by being verbal (not nominal) markers of reciprocity, by being markers of transitive (not intransitive) verb forms, or by displaying other morphological and/or lexical restrictions;
- Often various flexibilities in language use exist; speakers draw on different constructional means to express or implicate reciprocal semantics; in some contexts, reciprocal markers are obligatory, in others they are optional (e.g., English *They kissed* or *They met*, but optionally also *They kissed each other* or *They met each other*);
- Closely related languages often bear striking resemblance in the general form and function of linguistic material, which suggests a common historical origin (e.g., the reciprocal verbal suffix *-nyji* in Nunggbuyu and the reciprocal/reflexive circumfix *ma-V-nyji* in Nyulnyulan languages; chapter 5);
- Languages often rely on etymologically similar expressive means in order to encode reciprocal meanings (e.g., reciprocal markers derived from lexical items with meanings like 'meet,' 'fellow/companion,' 'exchange' or 'self'; see discussions on etymological sources in König & Kokutani, 2006, pp. 289–92; Evans, 2008, pp. 50–57; Heine & Miyashita, 2008, pp. 176–87).

Typologists commonly refer to these similarities in cross-linguistic perspective as *family resemblances*. Here are some examples from the literature: "The proper model for understanding this distribution, I would suggest, is [. . .] the family resemblance model, in which the fragments that end up as canonical instances are interrelated through partial formal-functional similarities" (Hopper, 2001, p. 125). "[R]ather we are dealing with family-resemblance phenomena: no two languages have any word classes that are exactly alike in morphosyntactic properties or range of meanings" (Evans & Levinson, 2009, p. 435). "Thus, we construe recurring patterns of language along the lines of Wittgenstein's (1953) notion of 'family resemblance': although there may be similarities between pairs of individual languages, there is no single set of features common to all languages" (Christiansen & Chater, 2008, p. 453). "And it is these other parts that produce the light 'family resemblance' that we see among the languages of the world, things that could indeed lead a Martian to think that they looked alike" (Everett, 2012, p. 201).

As is indicated by Christiansen and Chater (2008) above, the term "family resemblance" originates in Wittgenstein's (1953) *Philosophical Investigations*.

> Consider, for example, the activities that we call "games." I mean board-games, card-games, ball-games, athletic games, and so on. What is common to them

> all?—Don't say: "They *must* have something in common, or they would not be called 'games'"—but *look and see* whether there is anything common to all.—For if you look at them, you won't see something that is common to *all*, but similarities, affinities, and a whole series of them at that. [. . .] I can think of no better expression to characterize these similarities than "family resemblances"; for the various resemblances between members of a family—build, features, colour of eyes, gait, temperament, and so on and so forth—overlap and criss-cross in the same way.—And I shall say: "games" form a family. (Wittgenstein, 2009 [1953], Sections 66–67, p. 36[e])

Wittgenstein also applies this notion of games and their family resemblances to language, or better still, linguistic activity. "I shall call the whole, consisting of language and the activities into which it is woven, a 'language-game'" (Wittgenstein, 2009, Section 7, p. 8[e]). In everyday language, speakers use language in different contexts and for different purposes of behavior. They "game," to use Wittgenstein's term, with linguistic means to make their purposes understood by their interlocutors. "Doesn't someone who acts on the call 'Slab!' in such-and-such a way understand it?" (Wittgenstein, 2009, Section 6, p. 7[e]). The aptness of this characterization has been recognized by typologists throughout the latter half of the twentieth century. Examples like the marking of reciprocity demonstrate that natural languages do not converge on any single feature, either structural or functional, which could be posited as a cross-linguistically valid feature of *all* natural languages. The data sets describe complex continua of grammatical and/or functional features that are partially shared among languages around the world. The features of some languages display extensive overlaps in either form or function, while others only partially overlap, or not at all. Outliers at either end also form part of such continua and only indirectly resemble one another through partial overlaps among languages in between. Here is an example. In the Ngayarda language Yindjibarndi (West Australia), the reciprocal marker *-marri* is productive with various lexical items describing not only dynamic actions or states ('hit,' 'see,' 'speak') but also spatial arrangements (e.g., *ngarta-marri-* 'be side by side,' *purlu-marri-* 'be face to face'; Wordick, 1982, p. 85). In the Aslian language Mah Meri (Malaysia), by contrast, no reciprocal marking is productive with such stative forms. Based on data elicitation through video clips, Kruspe (2011) reports that "intransitive and stative verbs do not enter into reciprocal constructions, thus clips which depicted physical proximity, e.g., 'next to' (#8 people sitting next to each other), or 'leaning' (#35 sticks leaning together) in the stimuli set caused consultants great difficulty in responding at all" (p. 158). In the Australian language Yidiny (Northeast Australia), on the other hand, the suffix *-bany* is only productive as a marker of spatial juxtaposition (e.g., *walmbi:rr-bany* 'to be side by side,' *garba-garba:-bany* 'to be behind one another'; Dixon, 1977, p. 162; see chapter 6); it does not combine with dynamic verb forms. Hence, we observe a continuum from dynamic to static/spatial arrangements. Only one language, Yindjibarndi, displays a marking pattern spanning the whole range. The other two display patterns which overlap partially with the latter, but not between one another. The cross-linguistic similarity is thus one of family resemblance.

SUBSTANTIATING THE TOOL HYPOTHESIS

Following the thesis of this book, the phenomenon of family resemblances finds its explanatory basis in the tool concept that was introduced in chapter 3. The conception of language as "tool of communication" predicts precisely such structural and functional *similarities and differences* in cross-linguistic analysis. The variation is too diverse to explain it on the basis of innate categories specific to the acquisition of linguistic behavior. The only conceivable alternative from a cognitive, learning-oriented perspective is the acquisition based on more general, that is, context-dependent and problem-oriented learning abilities. As we have seen in chapter 3, these are already characteristic of tool-using practices in great apes. They already reach a degree of complexity in the prelinguistic domain which begins to resemble the intentionality otherwise only attributed to human behavior (Ingold, 1993; Malle et al., 2001; Hurford, 2007, chapters 1–3; Campbell, 2011; Dehaene, 2014).

At present, the typological literature does not recognize the tool hypothesis as a useful explanatory concept. The concept is not seen in connection with clear predictions that add to our knowledge about the cognitive and social foundations of language evolution and historical development. In chapter 3, we illustrated the basic characteristics of tool-using behavior in prelinguistic animals. Great apes are useful model organisms because their higher cognitive abilities are not yet connected with the ability to speak. They form cohesive, coordinated groups whose members depend upon one another for survival. Under such conditions, primatologists observe, great apes develop and conform to conspicuous group customs (Whiten et al., 1999; Van Schaik et al., 2003). These group customs vary unpredictably across groups, and these differences are not explicable either in terms of genetic or ecological factors. Hence, they give evidence to sociocultural transmission within the group (Whiten, 2005, 2017). These observations were spelled out at the end of chapter 3 and summarized as predictions for cultural variation in natural languages (*Predictions 1–4*). We tested these predictions for structural and functional variation in chapters 5 and 6, respectively, and found positive evidence for them. Hence, in respect to patterns of variation, we have already shown that the tool hypothesis makes faithful predictions in regard to language. It confirms the hypothesis that linguistic constructions are of *cultural origin.*

However, the tool hypothesis has still more to offer. We have so far only dealt with the *historical end products* of language use, not language use itself, as a dynamic process. In a next step, we identify various *context factors* that influence language use. These provide essential motivations driving language development. Just like the context factors in prelinguistic tool use, they provide clues that help explain the observed patterns of variation, including cross-linguistic similarities. In the literature, these factors are usually associated with *usage-based approaches* to language change (Kemmer & Barlow, 2000; von Mengden & Coussé, 2014; Ibbotson & Tomasello, 2016). We here associate them more specifically with the characteristics of *tool intelligence.*

Semantic Context Factors

The optional use or even unproductivity of reciprocal markers with certain verb forms (whose semantics nevertheless allow a reciprocal interpretation) is a conspicuous phenomenon across languages. For example, the unproductivity of the reciprocal marking pattern 'Rdp-V + *-(n)barriy*' in Dyirbal with verbs meaning 'fight' and 'divide' (Dixon, 1972, pp. 299–300; see chapter 6) is striking if we consider that other verbs quite similar in meaning (e.g., *baral-* 'punch, fight,' *wugal* 'give') must be marked by this formal pattern in reference to the reciprocal situation (e.g., *baral-baral-nbarriy* 'punch/fight each other'). Yet, a functionally oriented perspective, a tool perspective on language use that is, renders this phenomenon somewhat less exotic. It shifts the attention away from language form (marking as opposed to no marking) and toward the *subjects' intentions* in verbal exchange. Consequently, the *pragmatics* of language use, use in context, becomes the focus of attention and with it the question of how language users rely on linguistic means to achieve their goals under variable conditions of communication.

We thus assume that the use of reciprocal markers in language is a form of *communicative tool use.* We predict that the employment of reciprocal markers is primarily applied to contexts in which communicative challenges ("problem settings") arise. That is, just as one does not expect wild chimpanzees to use stone hammer and anvil whenever they crack open nuts, so one does not, as a functional linguist, expect language users to employ dedicated reciprocal markers in *all* contexts which afford a reciprocal interpretation. Rather, we predict that reciprocal markers are used as *tools* needed in some contexts of verbal exchange, but not in others. As a general tendency, we expect them to occur in contexts where the reciprocal event configuration is "hard to get," that is, in contexts where they function as dedicated tools which help extract the reciprocal meaning for an interlocutor to understand unambiguously. On the other hand, we predict that reciprocal markings occur less frequently in contexts where the reciprocal meaning is "easy to get," that is, in contexts where the reciprocal interpretation is obvious and recoverable from less dedicated means and/or other factors influencing speaker-hearer interaction.

Among functional linguists, the factor of *inherent (non-)symmetry* has long been an important aspect in the discussion of reciprocal marking patterns (König & Kokutani, 2006; Nedjalkov, 2007b; Haspelmath, 2007b; Evans, 2008).[1] Languages tend to be formally sensitive toward the semantic distinction between *non-symmetrical* (e.g., 'hit,' 'see,' 'give') and *symmetrical* (e.g., 'fight,' 'meet,' 'exchange') predicates (König & Kokutani, 2006, p. 272). This formal sensitivity shows in that verbs belonging to the former type of verb class (non-symmetrical) normally require explicit reciprocal marking, while the latter (symmetrical) do not always require it (see Figure 7.2). The explanation for this phenomenon draws on the pragmatics of language users. If someone uses or hears an expression like 'they fight' or 'they talk,' they will infer from context and world knowledge that mutuality might be implied. It doesn't have to, but it could be the meaning intended by a speaker. For when people fight, it takes at least two (A and B) and tends to go back and forth between them. A similar logic applies to other events like talking, agreeing, meeting, dispersing, resembling, and so on. With expressions like 'they hit' or 'they see,' on the other hand, pragmatics tend in another direction. If A hits B, B doesn't have to be hitting back. Hitting does not imply mutual/symmetrical involvement. The same is true of other events like seeing, helping, telling, and going past. In such contexts, trying to evoke the sense of mutuality costs more effort in verbal exchange. Counter-intuitions of interpretation have to be overcome. Hence the tendency of language users to invent more dedicated means (*verbal tools*) to express reciprocity under such more unfavorable conditions of communication. Yet, as this differential formal treatment of verb forms is assumed to emerge in the context of cultural-historical developments within each single speech community, languages are always expected to differ from one another in respect to the precise selection of verb forms treated in this way. Language users negotiate such meanings in context on a day-to-day basis and in various social and referential settings. As a result, the group of unmarked reciprocal verb forms in a language will always be language-specific, not fixed for all languages.

This prediction, based on both functional and tool-based considerations, holds true. In the Australian language Nunggubuyu, there are a number of inherently symmetrical verb forms such as *=wunda-* 'quarrel,' *=walmbalndha-* 'discuss,' *=libawa-* 'share,' *=rajara-* 'disperse,' *=yada-* 'connect,' and *=ngunyju-ga-* 'make equal' (Heath, 1981, 1984). Dyirbal, also an Australian language, only treats the verbs *bundalay* 'fight' and *munydjal* 'divide'

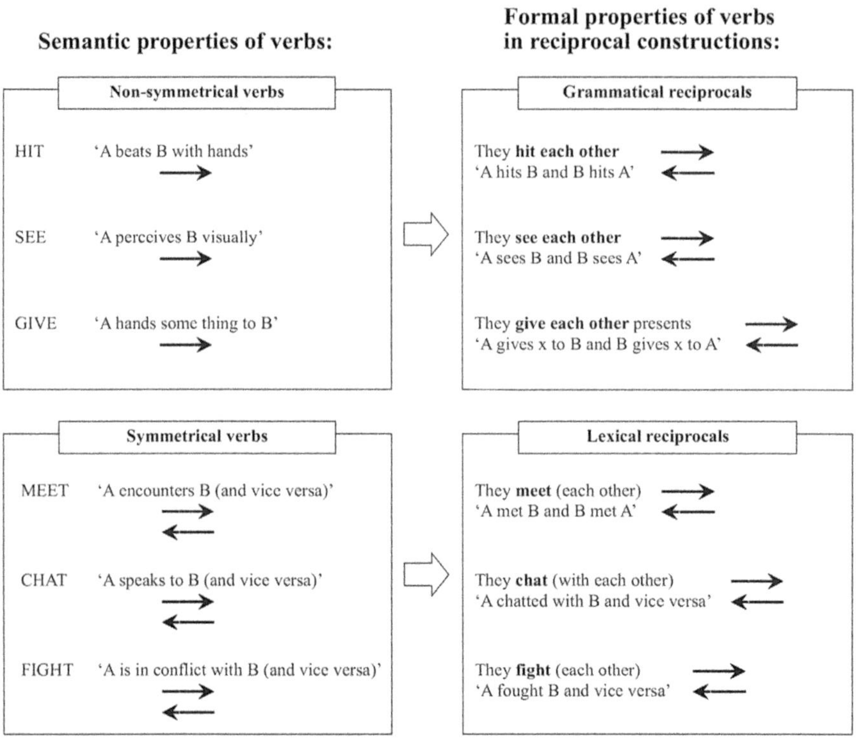

Figure 7.2 Reciprocal Marking of Symmetrical and Non-Symmetrical Verbs. The left half of this diagram depicts the inherent semantic properties (i.e., configurational meanings) of lexical verbs. The right half illustrates the formal marking that such verbs typically receive when used explicitly to denote reciprocal situations. At least in English, verbs with non-symmetrical event configurations (e.g., *hit, see*) need to be explicitly marked with the reciprocal expression *each other* (↔), while verbs with inherently symmetrical event configurations (e.g., *meet, fight*) need not (necessarily) be marked this way (indicated by brackets around *each other* in each example). Since non-symmetrical verbs need to be combined with additional linguistic material (here: English *each other*), the resulting reciprocal verb forms are referred to as *grammatical reciprocals*; since symmetrical verbs need not be combined with this expression in order to express the reciprocal meaning, the resulting verb forms are referred to as *lexical reciprocals* (Nedjalkov, 2007b). *Source:* Author created.

without special reciprocal marking (Dixon, 1972). In both Martuthunira and Panyjima, it is the form *patharri-* 'fight' which behaves in this way (Dench, 1991, 1995), while in the closely related Jiwarli it is not the verb *patha-* 'hit, fight' but the form *wangkaarni-* 'speak, talk' which may be used without the reciprocal marker (Austin, 1997). The typological literature provides further examples from other language families. For instance, the Uralic-Yukaghir language Kolyma Yukaghir allows the symmetrical verbs *kimd'i:-* 'to fight' and *šaqal'e-* 'to gather' to be used with or without explicit reciprocal marking (Maslova, 2007b, pp. 1853–54). The Aslian language Mah Meri (Malaysia) treats the following verbs in a so-called bare conjunct construction: *kahɔn* 'to kiss', *səroh* 'to meet,' *bwal* 'to converse,' *salạp* 'to greet,' *pəloʔ* 'to hug,' *ŋaheʔ* 'to quarrel, argue, fight,' *tuka (barạk)* 'to exchange (things)' (Kruspe, 2011, p. 152). Finally, English, too, has various lexical items that are frequently used as lexical reciprocals (Nedjalkov, 2007b; Knjazev, 2007), as is illustrated by the following example(s): *They met / dated / danced / emb raced / kissed / agreed / married / quarrelled / split up / divorced* (König & Kokutani, 2006, p. 274). The precise lists of lexical items treated as lexical reciprocals vary from language to language; only the underlying pragmatics motivating such instances are the same.

Moreover, typologists note that languages differ greatly with respect to the degree of flexibility allowed in language use. In some languages, the use of reciprocal marking is highly flexible and depending on whether speakers choose to use a reciprocal marker or not. For instance, in Kolyma Yukaghir, the above-mentioned lexical forms ('fight,' 'gather') also (optionally) combine with the explicit reciprocal marker *n'e-* to form complex reciprocals (*n'e-kimd'i:-* 'to fight with each other'; *n'e-šaqal'e-* 'to gather with each other'). In this context, Maslova observes: "The intention to highlight the reciprocal nature of a referential situation," seems to play an important role and thus, given this intention, "results in adding the morphological reciprocal marker to a lexical reciprocal" (Maslova, 2007b, p. 1854). This flexibility is thus motivated by context and intention, not linguistic convention. The same is true of Mah Meri, though speakers here rely on quite different motivations. Kruspe (2011), who relies on the video clip material designed by Evans et al. (2004) (see Evans et al., 2011b), observes: "'Naturally' reciprocal verbs were also frequently used in the double distributive construction [. . .] to express less prototypical cases of reciprocity, such as chain sequential situations (#14, #28), adjacent simultaneous situations (#19, #61), and pairwise sequential situations like #30 [. . .], providing elaboration of the prototypically reciprocal sub-event" (p. 159).

In other languages, such as the above-mentioned Dyirbal and Martuthunira, such flexibility does not exist with symmetrical predicates; they are either explicitly marked or unmarked by convention (though perhaps even they would depart from such conventions if consultants were exposed to similar conditions as devised by Evans et al., 2004). This obligatoriness of formal marking is by no means always connected with the inherent (non-) symmetry factor. It may also have to do with avoiding *ambiguity*. This is exemplified by languages with polysemous marking patterns, that is, languages whose reciprocal markers also function as nonreciprocal markers like markers of reflexive ('hit oneself') or collective ('laugh together') action (see chapter 6). In such cases, it sometimes happens in language use that utterances become ambiguous between two or more readings/interpretations (e.g., 'they fight each other' as opposed to 'they fight together (against)'; or 'they hit themselves' as opposed to 'they hit each other'). Such potential ambiguity exists but is often resolved implicitly through context (common ground; previous discourse). Yet, language users do not always allow such potential ambiguity to persist. They invent additional means to make their intended meanings explicit to the interlocutor. Thus, in German, the reflexive/reciprocal pronoun *sich* is often either combined with the intensifying expression *selbst* to express the reflexive meaning (*sich selbst schlagen* 'to hit oneself'), or else with the reciprocal adverb *gegenseitig* to express the reciprocal meaning unambiguously (*sich gegenseitig schlagen* 'to hit one another') (König & Kokutani, 2006, p. 284; see also König & Vezzosi, 2004). In German, though, such additions are still optional. In other languages like Bolivian Quechua, the situation is different. Here, the additional verbal prefix *na-* is mandatory in combination with the reflexive element *-ku-* if the reciprocal meaning is intended (Muysken, 1981, p. 454, p. 464; cited from Maslova, 2008, p. 250). In other words, in this language, the disambiguating element has become part of the linguistic convention and no longer depends on just pragmatic considerations—although originally, the situation might have been similar to present-day German.

Many more examples could be given here; however, the basic point has been made. Reciprocal marking patterns in natural languages show conspicuous asymmetries in the formal marking of reciprocity. These asymmetries can be motivated by pragmatic factors that guide speakers' linguistic activity. These relate to, for example, real-world knowledge about the events referred to by lexical items and the inferences speakers and addressees make about them when used in any given context of speech. "Hard-to-get" reciprocal meanings are marked more explicitly than "easy-to-get" reciprocal meanings as an adaptive strategy to retain expressiveness even under unfavorable conditions of usage. Yet the degree of subsequent conventionalization of such differential pragmatics remains language-specific, that is, bound to historical accident. This situation in natural languages is analogous to prelinguistic tool uses (see chapter 3). Tool practices like the use of stone hammer and anvil are also sensitive to ecological factors (e.g., hard-to-get versus easy-to-get food resources). Yet, their exact distributions still vary across groups and thus always remain products of sociocultural transmission.[2]

Structural Context Factors

We resume the tool perspective in respect to structural limitations in language use. Reciprocal markers in different languages are subject to similar yet also different structural restrictions. Some markers are only productive with verbal expressions (e.g., Nunggubuyu *-nyji*, Heath, 1984; Yimas *tɨ-*, Foley, 1991); in other languages, they also productively combine with nominal expressions (e.g., Fijian *vei-*, Dixon, 1988). Some are only productive with transitive verbs (e.g., Dyirbal *-(n)barriy*, Dixon, 1972; Duungidjawu *-ye/-Cmdje*, Kite & Wurm, 2004), others are also productive with intransitive verbs (e.g., Dyabugay *-lndjirri/-nydjirri*, Patz, 1991; Martuthunira *-marri*, Dench, 1995). Some marking patterns are only productive with simple verbs (e.g., *ma-V-nyji* in Nyulnyulan languages, McGregor, 1999); others are also productive with compound verbs (e.g., Kâte Rdp-V + *e-*, Pilhofer, 1969). Moreover, in some languages, reciprocal markers are even sensitive to the syllabic structure of verb stems. In the Aslian language Jahai (see Map 4.1, p. 68), for instance, the reciprocal marker (/Ca/, <*a*>) is only productive with monosyllabic and sesquisyllabic ('one and a half'-syllabic) verbs. "The reciprocal morpheme has not been documented in connection with disyllabic verbs" (Burenhult, 2011, p. 166).

It is of course not possible to explain these various structural limitations under a single motivating factor. Various factors are responsible for such different distributional outcomes. Before we consider some of them, however, we recall the tool concept. The tool perspective offers a fresh view on such linguistic phenomena. It helps explain why they are still consistent with a functionalist, usage-based account of language development. It again reshifts the focus from purely formal to pragmatic and historical considerations. The term *pragmatic* is here used in an unconventional way since normally it is only understood in connection with the meaning of linguistic utterances. Nevertheless, the tool concept dismisses the dichotomy (form–meaning) and also considers the handling of linguistic form under the influence of pragmatic factors. Tools are "constructed" means that are invented and recruited in context (historically). A stick can never be used if it is not combined with various other manipulative units (grasping; twisting; poking; etc.). In some cases, it even requires the combination with other materials (e.g., another stick; leaves; a stone). This context-dependency and "construction" of tools thus always accompanies their usage and persists over subsequent trials. Furthermore, its innovation and transfer is always embedded in other conventions already established within the group. Thus, wild-living chimpanzee populations often show conspicuous preferences for certain food resources (e.g., palm trees) and invent various techniques to exploit them in different ways. In some populations, such techniques remain limited to only such preferred food resources, although from a purely logical point of view, they could easily be extended to elsewhere. Similarly, some populations develop clear preferences for certain objects-as-tools, such as leaves or sticks. Other materials are equally available to them; however, they do not favor them as tools. Such preferences are not innate; they are preferences born out of social customs already existing within the group. Such customs persist across the generations and thus give evidence to deep-historical roots (Boesch, 2012, 2013). These roots form the backbone for any changes occurring under extant conditions of behavior.

In that respect, natural languages are not that different. It is often possible to align restrictions in the productivity of reciprocal markers with more general patterns (restrictions) of grammatical marking in a language. These in turn may have a semantically motivated historical origin. For example, Dixon (1972, 1977) points out for languages such as Yidiny and Dyirbal that they show particular sensitivity to the distinction in formal treatment between transitive and intransitive verb stems, which is pervasive throughout the grammar of these languages. The fact that the reciprocal marking pattern in Dyirbal is strictly restricted to transitive verb stems is thus not only a specific feature of reciprocal markers but also typical of other marking patterns in the language (e.g., the use of the reflexive suffix). The Ngayarda languages of West Australia, by contrast, have undergone a drastic development in the opposite direction, which has resulted in a less marked distinction between the formal treatment of transitive (semantically other-directed) and intransitive (semantically non-directed) verbs.

> These languages have changed their basic case marking pattern from ergative/absolutive to nominative/accusative and have developed an active/passive voice distinction. It is argued that the modern accusative patterns developed through the generalizing of a semantically motivated intransitivising construction as the common vehicle for transitive verb stems. The accusative case marking pattern then facilitated the development of a productive passive construction. (Dench, 1982, p. 43)

This loss of a formal distinction between transitive/intransitive shows, for example, through a lack of case-marking distinctions between transitive and intransitive subjects (nominative-accusative alignment), and the frequent possibility to omit the overt expression of arguments in object position (argument ellipsis). Along these lines, Dench (1987) remarks that "unlike most Australian languages, those of the Ngayarda group have evolved a nominative/accusative case-marking system [. . .], a change which has resulted in a reduction in the importance of transitivity in these languages" (Dench, 1987, p. 325). In regard to the verbal *-marri* suffix, he then continues to argue: "The loss of a strong transitivity contrast has meant that the reciprocal suffix has no important part to play in syntactic organization, and this freedom has allowed the extension of the suffix function into an area largely unaffected by syntactic phenomena, the marking of kinship relationships" (Dench, 1987, pp. 325–26). Although Dench makes these considerations specifically in regard to the "kinship" function of the collective suffix (see chapter 6), it seems possible to also extend this consideration to the fact that the verbal *-marri* suffix became established as both reciprocal/collective marker in the first place, becoming productive with both transitive and intransitive verbs.

The Nyulnyulan languages in the Kimberley region of Northwestern Australia, in turn, illustrate that special organizations in the area of the lexicon of a language can have an important impact on the development of reciprocal marking. As is discussed in McGregor (2002) and elsewhere, it is a special feature of only some languages in (mostly) Northwestern Australia to have both simple inflecting verbs and uninflecting verbs (*preverbs*, *coverbs*), the latter of which are used in construction with a small set of simple inflecting verbs in compound verb constructions (McGregor, 2002, pp. 25–26, pp. 107–14; see also Dixon, 2002, pp. 183–87). As was shown in chapter 5 for Nyulnyulan languages, the structural distinction in the expression of predicational meanings (simple versus compound) goes hand in hand with a formal distinction in reciprocal marking. The use of the verbal circumfix *ma-V-nyji* is only productive with simple inflecting verbs; the use of the auxiliary *-banji/-wanji* is only productive with preverbal elements. In the distantly related language Nunggubuyu, by comparison, such a formal distinction between simple verbs and preverbs does not exist in the formal architecture of the lexicon. Neither is the verbal reciprocal suffix *-nyji* restricted in any way similar to the Nyulnyulan *ma-V-nyji* pattern. Interesting is again the consideration of semantic-pragmatic factors. Hence, it is a conspicuous fact that the simple inflecting verbs which combine with preverbal elements to form compound constructions do not allow reciprocal marking by *ma-V-nyji* even if used on their own. According to McGregor (1999, p. 96), this is due to the fact that the lexical meanings of these verbs do not lend themselves for the expression of reciprocal event types. They do not yield sensible reciprocal meanings, at least not in the minds of language users. This, in turn, may have prohibited speakers to associate the circumfixal marking pattern with these simple verbs even if used in construction with additional preverbal elements that would allow sensible reciprocal interpretations. The natural consequence would have been to invent a new way of expressing reciprocity with preverbal elements. The recruitment of the inherently symmetrical verb *-banji/-wanji* 'exchange' seems to have been the solution to the problem.

Finally, we return to the Aslian language Jahai on the Malay Peninsula. As we mentioned above, the reciprocal marker is here only productive with verbs of a particular syllabic structure. The marking consists of infixation and in some cases also partial reduplication (Burenhult, 2011, p. 166; note that in the ensuing description, we closely follow this reference). With monosyllabic CVC bases, the marker consists of a syllable /Ca/, with an underspecified consonant onset and a prespecified vowel *-a-* which is prefixed to the verb base. The consonant slot of the affix is filled by a copy of the onset of the CVC base. Resulting reciprocal formations are, for example, *ca-cɔl* 'to tell each other' (from *cɔl* 'to tell') or *ʔa-ʔɛl* 'to look at each other' (from *ʔɛl* 'to look'). With sesquisyllabic ('one and a half'-syllabic) verb bases, on the other hand, there is no partial reduplication, only infixation of *<a>* before the final CVC syllable of the base, producing reciprocal forms like *s<a>mɛɲ* 'to ask one another' (from *smɛɲ* 'to ask') or *b<a>dil* 'to shoot one another' (from *bdil* 'to shoot'). With disyllabic verb bases, however, it is not productive at all (Burenhult, 2011, p. 166). The same principle applies to another reciprocal marking pattern which Burenhult (2011) refers to as *distributive*, though for our present purposes we do not need to enter into the details of the construction.

Several considerations are of particular interest with respect to the data on Jahai. One is that speakers of this language obviously adhere to very special rules of phonological, syllabic verb structure building which clearly differ from other languages. Especially from a Eurocentric perspective, it is a conspicuous feature of the language.[3] Hence, language users grow up to acquire a distinct "constructional rhythm" of what verb forms are composed of in everyday usage, and this acquired knowledge differs dramatically from speakers' intuitions in other languages around the world (e.g., English or German). It involves a heightened degree of sensitivity of syllable structure and a sense of "saturation" (completeness) of word forms once they assume a certain degree of complexity. We can observe the following tendency: short (monosyllabic) forms are modified until they are disyllabic; medium (sesquisyllabic) forms are also modified, yet less materially, until they too are disyllabic; while verb bases that are already disyllabic to begin with, are not modified at all, but left unmarked. This tendency to modify verbs until they reach a common syllable structure seems to pose a general "organizing principle" which language users, unconsciously and by routine, adhere to in everyday usage. It has become a sorting strategy to organize linguistic material and thus, implicitly, also meaning in this language.[4]

Yet, this is not all. "Jahai shows evidence of substantial lexical borrowing from Malay, the Austronesian majority language of the peninsula" (Burenhult, 2011, p. 163). These Malay loans are mostly di- or trisyllabic, unlike most indigenous verbs of Jahai (Burenhult, 2011, p. 168). Interestingly, in the video-clip tests developed by Evans et al. (2004), this divide within the lexicon also plays a decisive role in the expression of reciprocity. If speakers choose Malay loan verbs to express particular events, they tend not to use any reciprocal marking at all. It appears that they merely describe the events in nonreciprocal terms, some of their utterances perhaps merely implicating reciprocity, or not at all. If, on the other hand, the informants choose to express reciprocity, they rely on indigenous Jahai verbs, to which reciprocal marking is applicable. Only some speakers on occasion deviate from the standard convention by using disyllabic Malay verbs with the *distributive* ("reciprocal") morphology:

> There are occasional exceptions to the restriction. On one occasion one consultant spontaneously used the distributive form *h<it>mat*, from the disyllabic Malay loan *hamat* 'to chase' (Malay *hambat*). Another consultant used the distributive form *tiʔmoʔ*, from *tumoʔ* 'to hit' (Malay *tumbuk*). And two consultants accepted a distributive derivation *sl<it>mat* of the Malay loan *slamat* 'to greet' (Malay *selamat*). (Burenhult, 2011, p. 168)

What these exceptions suggest is that under certain conditions of usage, speakers of the language become inventive and start extending the general principle of marking reciprocity to novel lexical items not usually subsumed under their convention. Disyllabic Malay loans start receiving the same morphosyntactic treatment as native words with less syllabic complexity. This in turn is indicative of another motivating factor currently at work in the larger speech community. The incorporation of Malay loans into everyday usage. This is a social factor, a typical symptom of language contact, which is sure to be connected to various sociopolitical dynamics at work within the larger community. Given that this pressure of incorporating Malay loans into daily discourse exists, it would be a natural (adaptive) reflex to also start extending existing marking patterns to these novel forms. Perhaps only in some contexts, that is, piecemeal, not in others. The particular historical setting might thus currently favor the gradual overturning of a long-lived linguistic convention. Morphosyntactic marking patterns grow less sensitive to syllabic verb structure and thereby become gradually more productive means of marking reciprocal (and distributive) meanings.

A somewhat related observation was made with regard to the Australian language Nunggubuyu (see chapter 6). We observed, based on Heath (1984), that the reciprocal suffix *-nyji* was only marginally productive as reflexive marker with a handful of monosyllabic verbs. The otherwise standard way of expressing reflexivity in this language, namely by means of the verbal suffix *-i*, does not apply in these instances. To repeat Heath (1984): "In general, mono-syllabic verb roots either do not have or make little use of Refl derivations" (p. 391). This might thus be another example where native speakers of a language have developed basic intuitions of speech that include an intuition about either maximal or minimal syllabic complexity in the articulation of complex (inflected) verb forms. Such intuitions may permeate various areas of complex word formation and thus influence the choice of morphological means that are considered most appropriate for the marking of specific (configurational) meanings. What is interesting about such phenomena is the fact that the choices of linguistic means ("tools") are here directly motivated by the "physical handling" of such devices in everyday speech events (i.e., patterns of articulation that are regular and thus economical), and not by the meanings expressed. Under the tool perspective, it is quite natural to view such factors as essentially *pragmatic* and *usage-oriented.*

To conclude, the various examples illustrate that linguistic means, such as reciprocal markers, are by no means only sensitive to the lexical meanings of forms. They also have structural properties and thus are sensitive to the "constructional rules" of a language system. Given the tool analogy, this is hardly surprising. It is precisely the characteristic we also associate with the making of tools. Tools are "constructed" items and they only function in conjunction with other materials in order to form functional wholes. Once tool use reaches higher levels of complexity, the various items begin to enter into constructional relationships between one another. The structural and functional fitting between these various items poses problems in their own right, disconnected from the final goals which they are intended to serve. The fitting between two sticks, which are inserted into one another, is not directly connected to the fishing of termites. It ultimately serves the purpose of fishing termites; but stick insertion is a constructional problem not connected to the fishing action itself. The strategies of dealing with such additional constructional challenges are equally diverse as in the case of simpler tool uses. Different solutions can be found. Yet, strategies that prove successful under the conditions of usage and spread within the group are likely to persist as group-specific trends. Moreover, they are also likely to inform problem solving in other domains of behavior. The results are general deep-historical trends of development that are characteristic of some cultures but not others. Only special historical events are likely to overturn such trends and may lead to reorganizations that follow new structural principles of behavior.

We can see the analogy to reciprocal marking patterns in natural languages. The structural restrictions of reciprocal markers are often bound up with general features of syntactic, morphological, and phonological organization in individual languages. For some of these, it is even still possible to conjecture motivating factors of language use that are held responsible for the historical development of these structural design features in the first place. For others, it is not possible under extant conditions, due to the opaqueness of their historical origin.

PRELIMINARY CONCLUSION

The predictions made by the tool hypothesis correctly anticipate patterns of variation in cross-linguistic comparison. Functional linguists commonly characterize partial overlaps between these patterns of variation as family resemblances. According to the hypothesis, they are the cumulative result of historical-cultural developments within individual speech communities, which are synchronically driven by various pragmatically motivated factors involved in language use. Depending on the historical constellations, the same motivating factors may drive linguistic activity in different speech communities. The causality of these context factors is proven by predictable patterns of distribution that yield similar outcomes in different languages. The predictions also acknowledge the historical-cultural dimension of the process, which implies different chains of events and therefore unpredictably different patterns of distribution cross-linguistically. In other words, the details of such distributional patterns always vary from one language to another. In the present study, the cognitive substrate responsible for negotiating such goal- and context-oriented forms of language use has thus been identified as essentially the same cognitive substrate ("tool intelligence") on the basis of which great apes perform complex forms of prelinguistic manipulative behavior ("tool use") and develop social customs ("culture").

NOTES

1. In the literature, different labels have been introduced (and proposed) for this class of verbs: *naturally reciprocal verbs* (Kemmer, 1993), *reciprocal verbs* (Siloni, 2008), *symmetric(al) predicates* (König & Kokutani, 2006; Evans, 2008; Dimitriadis, 2008), or *allelic predicates* (Haspelmath, 2007b). We here mainly adopt König and Kokutani's and Evans' terminology, though with a narrower focus on "verbs," not "predicates" more generally.

2. Within the functionalist framework, the above considerations are typically classified as considerations of "iconicity" (Haiman, 1983, 2000; Givón, 1985, 1991; see review in Haspelmath, 2008b). In works by Haspelmath and others, at least some of these iconicity factors are contrasted with other explanatory factors, which are conceived of as "frequency effects" or "processing constraints" (Hawkins, 2004; Haspelmath, 2008a, 2008b). As a result of this debate, the range of possible motivating factors for the development of morphosyntactic asymmetries in cross-linguistic perspective has become more complex and differentiated even within the functionalist debate. Different usage-oriented factors are considered as alternative driving forces of language change and, in particular, processes of grammaticalization. The present investigation does not exclude the influence of frequency effects on the emergence of structural dependencies in language use that define grammaticalization. However, given the tool analogy, a more primary question appears to be why certain structural/functional complexities arise in language use in the first place, irrespective of any frequency-of-usage effects that may unfold in everyday discourse. This primacy of conjecture arises from basic considerations of tool use practices as "problem-oriented" behaviors. The question of how under certain conditions of goal-oriented

behavior material objects turn into tools, that is, intentionally handled, necessary devices to solve problems, seems more basic than the question of how such variants of tool-using behavior evolve once they enter social group customs and become innovated through frequent repetitions of behavior in specific contexts of usage. The above discussion thus seems to be more closely associated with considerations of iconicity than with considerations of frequency within the functionalist debate.

3. We must qualify this claim, however. The Eurocentric perspective allows us to consider the Jahai phenomena as something "exotic" for which we have no *direct* equivalents in European languages. However, we do not have to look far to find similar, comparable phenomena. For instance, the consideration of syllabic structure (complexity) of words does play a role in discussions concerned with word order in French noun phrases (long versus short adjectives that either precede or follow the modified noun; e.g., *coïncidence heureuse* versus *grande coïncidence*). Or similarly, it has been invoked in the discussion on synthetic as opposed to periphrastic comparative formations in English (e.g., *longer* versus **more long* as opposed to *more interesting* versus **interestinger*).

4. For a description of this fact in more linguistic terms, see Burenhult (2011, pp. 167–68).

Chapter 8

The Relationship to Biological Concepts of Reciprocity

In this chapter, before concluding (chapter 9), we cast yet another perspective on the discussion of reciprocity. We revisit some of the language data with yet another question in mind. What are the semantics/pragmatics that reciprocal markers express? In other words, what is the concept explicitly encoded by reciprocal markers? Are they markers of social reciprocity?

Over the past one and a half decades, since the early 2000s, linguists with a functionalist (language-typological) background have repeatedly made speculations as to a possible connection between the semantic encoding of reciprocity in language and social cognition underlying flexible, cooperative forms of behavior among humans (König, 2005, 2011; König & Gast, 2008b; Evans et al., 2011b; Majid et al., 2011a). Discussions about the phylogenetic origins and ontogenetic substrates of cooperative behaviors of humans (and other animals) have a long-lasting tradition in various disciplines outside of linguistics, among them behavioral biology, evolutionary psychology, and anthropology.[1] Notwithstanding the implications of such nonlinguistic theories of social behavior, functionalist linguists have speculated that the importance of social reciprocity among humans also shows in language use in that languages employ dedicated means for the expression of reciprocity. Though the same linguists have been cautious and self-critical in drawing such connections, it has nevertheless become a standing hypothesis in the field to claim that such a connection between social cognition and linguistic behavior exists:

> What I have tried to do in this talk [paper] is to initiate a dialogue between linguistics and all those fields in the humanities and social sciences which are interested in concepts of reciprocity and their encoding in the languages of the world. [. . .] Constructions specialized for the expression of reciprocity are found in all languages. The development of such constructions is fed by several sources and it is probably a reflection of the importance of "reciprocity" for human interaction

> and survival that languages typically display several strategies for encoding reciprocity, which are not necessarily synonymous. (König, 2005, pp. 18–19)

> Reciprocity lies at the heart of human social life. [. . .] The general consensus is that cooperation and reciprocal exchange could only have evolved in the context of group selection [. . .] if the group was a culture bearing unit [. . .]. Early in the course of human prehistory, over a million years ago when the first cultural traditions of tool manufacture are manifest and thus when we first took the cultural turn as our special mode of adaptation, the predisposition to sharing and reciprocity would have had to have been in place. If this is correct, reciprocity lies deep in our mental makeup, and one would expect to find many manifestations of it in human psychology and behavioral predisposition. A straightforward prediction is that concepts of reciprocity should be reflected universally in human cognition and language. There ought to be a universal core concept, and its expression in language should be pervasive. But, given that the whole point of culture is to provide a novel means of fast adaptation to different ecological niches, the cultural manifestation of these underlying concepts could be rather diverse (Evans and Levinson, 2009). (Majid et al., 2011a, p. 1)

The precise implications of such statements are not clear unless we draw a narrower distinction between a weak and a strong interpretation of terms. The *strong interpretation* implies a close link between linguistic and nonlinguistic concepts of reciprocity. In this view, the reciprocal construction is a communicative device specifically functionalized for the expression of socially reciprocal behavior. This in turn implies that the invention and subsequent development (conventionalization) of reciprocal marking patterns in language is specifically motivated by those cognitive assets also underlying nonlinguistic social forms of reciprocal action. The *weak interpretation*, by contrast, states that reciprocal constructions are not functionally dedicated means for the expression of socially cooperative (reciprocally altruistic) behaviors. Instead, they are merely dedicated to the expression of symmetrical/mutual relations of whatever kind, whether these be social or nonsocial. In this view, the only aspect linking linguistic reciprocity to social forms of reciprocity is the natural tendency in humans to be preoccupied with the social affairs of daily life. This would explain the tendency in language users to employ reciprocal constructions most frequently in reference to social (as opposed to nonsocial) encounters between humans. In particular, it is the strong interpretation at stake in the discussion below.

ARGUMENTS *FOR* A CONNECTION

Considerations in favor of a connection are of the following kind: the frequent employment of reciprocals in reference to social interactions between humans; the reliance on culture-specific notions of social reciprocity for the interpretation of reciprocals; the employment of reciprocals in reference to non-symmetric event configurations. We exemplify each point below.

Reference to Social Interactions

Reciprocal markers frequently occur in construction with (verbal) predications denoting social interactions between two or more human individuals. Most commonly, such interactions are set either in the domain of antagonistic action (fighting, spearing, hitting, pushing, quarrelling, insulting, etc.) or joint action (talking, looking, sending, painting, helping, giving, etc.).

> In our daily lives, all sorts of interactions and escalations in social behaviour involve activities that require the reciprocally-directed activities of both parties. In fact, reciprocal activities occupy much of the narratives that interest us in novels, films and soap operas, from courtship through marriage to divorce, from meeting through exchanging gifts or gossip to fighting. [new paragraph] To talk and reason about such activities and principles in any accurate way we need linguistic devices to represent them. The majority of languages include special expressions for representing reciprocity. (Evans et al., 2011b, p. 2)

Indeed, it is fair to say that if we want to identify a dedicated reciprocal construction in a language, we first look for one in reference to social interactions in one of these areas of life (meeting, exchange of favors, gossip, fighting, etc.).

A narrower understanding of social reciprocity focuses on the exchange of favors and the maximization of individual gain and loss accounts, including the establishment of trust and collaboration within human societies (Mauss, 1990 [1924]; Hamilton, 1964; Trivers, 1971; Axelrod, 1984; Berg et al., 1995; Stegbauer, 2010 [2002]; Fehr & Fischbacher, 2004; De Waal, 2005a, 2005b; Moll & Tomasello, 2007; Tomasello, 2009; Molm, 2010; Cook et al., 2013). The situational prototypes of this type of reciprocity revolve around cooperative interactions in the domain of 'exchange,' 'sharing,' 'hospitality,' 'helping,' and 'support' (König, 2005, p. 5), but just as well include negations of cooperation in domains like 'warfare,' 'revenge,' 'competition,' 'abuse,' or 'cheating.' Turning to language use, we do indeed find that language users employ their reciprocal constructions in reference to such exchange relations in the narrower sense. Moreover, the reciprocal formations sometimes carry idiomatic and culture-specific meanings, whose literal meanings are drawn from concrete settings of daily life. For example, in the Papuan language Kâte (see Table 8.1), a notion like 'make peace' is drawn from the action of 'chewing betel nut together,' which is apparently a distinct sign of abolished hostility among the Kâte people. Likewise, expressions of 'mutual assistance' are taken

Table 8.1 Reciprocal Verbs Expressing "Social Cooperation" and "Social Competition" in the Papuan Language Kâte (see Map 4.1, p. 68). For detailed references of table entries see Table 12.5 in Appendix C.

Concept	*Reciprocal verb*	*Translation*
'exchange'	*boleŋ babac ezo*	'to exchange valuables to make peace by mutual paying of the wergild'
'make peace'	*huc nâṅâc ezo*	'to chew betel with each other, as sign for abolished hostility and restored peace'
'assist; unite'	*ʒiʒimic ezo*	'to assist each other in closing the cracks of the canoe; fig.: to help/assist each other; to stand by each other; to unite (against enemies)'
'help'	*ŋiŋiec ezo*	'to help each other fill the nets; fig.: to help each other'
'make war'	*qaqac ezo*	'to beat each other, to fight with each other, to make war'
'take revenge'	*okac babac ezo*	'to take revenge on one another'
'outdo'	*fefec ezo*	'fig.: to vie with each other, to upraid each other, to outdo each other in abusing'

Source: Author created.

from practical daily activities requiring mutual coordination and assistance like 'filling nets' or 'fixing a canoe' (see entries in Table 8.1).

Another intriguing example are the social kinship markers in Australian Aboriginal languages.[2] Dyirbal and Ngiyambaa are cases in point (see Map 4.1, p. 68). In both languages, kinship markers take the form of nominal suffixes, which are attached at the end of a noun denoting some social kinship relation. In Dyirbal, for example, the nominal expression *ŋagi* 'grandfather, grandchild' is suffixed with the nominal suffix *-djirr* and thereby derives a reciprocal noun *ŋagidjirr* 'grandfather and grandchild together' (see example [24]). Likewise, in Ngiyambaa, the nominal suffix *-galaydja:N-* attaches to a number of nominal expressions to denote mutual kinship relations between two (or more) people (see example [25]).[3]

(24) DYIRBAL

bayindayi	*ŋagi-**djirr***	*baninyu*
there.Nom.I.short distance uphill	*ŋagi*-Recip	come.Pres

'Two people, one of whom is maternal grandfather of the other, a short way uphill are coming.' (Dixon, 1972, p. 234)

(25) NGIYAMBAA

manabi-nji=li:-na	*ga:dhi:ŋ-**galaydja:***
hunt-Pst=1.Nom.Du-Excl	sister-Recip.Pl.Abs

'The two of us sisters were hunting.' (Donaldson, 1980, T8.1, p. 325)

It is well known that in Aboriginal culture the social interactions between people are strongly organized around the social kinship relations between them (Dench, 1987; Dixon, 2002). Dixon (2002) explains:

> The Australian Aborigines never developed agriculture. Like almost all hunter-gatherer communities across the world, there is no chief and no set of stratified social classes. Everyone in a tribe has specific social obligations towards everyone else, according to a finely articulated classificatory kinship system. [. . .] There is (or was) a classificatory kinship system, with every person in a community related to every other through a series of mathematical-like rules of equivalence. Each Australian community has strict conventions of how one should behave with each class of relatives. (Dixon, 2002, pp. 3–4)

It is therefore plausible to assume that the *linguistic encoding* of such social kinship relations through kinship terms as well as nominal kinship markers is a direct consequence of their sociocultural importance within the community. It is less clear, however, whether this finding gives direct support to the strong interpretation tested here. To be sure, the existence of rich repertoires of kinship terms in Aboriginal languages indicates a preoccupation with such relationships within the respective communities. Their various meanings entail complex social relations. Different languages around the world possess different repertoires of social kinship terms, highlighting their correlation with different systems of social organization and mutual dependency (Lynch, 1998, pp. 251–56). However, all this is not proof of the more specific hypothesis that linguistic reciprocity entails social reciprocity. For reasons to be explored further below, it is still conceivable that reciprocal markers, even nominal kinship markers, express mutuality on a psychologically far more detached or simply *different* level of perception.

In this context, it is perhaps useful to recall the primary functions and foundations of language use. Linguistic devices are relied upon by speakers with the purpose of directing the attentions of others. Following Kolodny and Edelman (2018), "language constitutes a toolkit that effectively supports an individual in influencing the state and the behaviour of others" (p. 3). Arguably, this purpose to communicate, as a way to influence others, is not sustained by social emotions; rather, it is sustained by pragmatic conjectures of what another person may and may not perceive or understand in a given context of interaction. It is based on an intentionality that is aimed at directing the attention of others "in particular ways so that they will do, know, or feel what the speakers want them to" (Tomasello, 2008, p. 343). Linguistic means are thus *subjective*, *intentional*, *manipulative* devices ("tools") used to sketch real-world experience in extremely crude, schematic ways such that other persons can guess the configuration of objects and events that are important to another person (the speaker) in a specific context of behavior.

The social or other values attributed to these real-world scenarios is mostly a function of experience, not the linguistic means used to evoke them. In Daniel Everett's terms, linguistic means are greatly "underdetermined" in meaning and based on "dark matter" knowledge that emerges through a complex interplay between context, culture, and human intelligence (Everett, 2016, p. 1, p. 26, pp. 178–93; 2017, pp. 3–5).[4] As we will discuss below, the crudeness of reciprocal schemas supports such an analysis and thereby favors the weak over the strong interpretation as regards the connection to social reciprocity.

Reference to Generalized Reciprocity

Second, it is pointed out by some authors that even concepts of "generalized reciprocity," as opposed to "direct reciprocity," are encoded by reciprocal constructions (König, 2005; Evans et al., 2011b, p. 5). Thus, sentences like English *Different generations should help each other* or *Immigrants from Turkey should help each other* do not have to imply that 'generation/immigrant 1 should help generation/immigrant 2 and vice versa' (direct reciprocity) but may as well mean that 'generation/immigrant 1 should help generation/immigrant 2 which/who in turn should help generation/immigrant 3, and so on' (generalized reciprocity). The latter interpretation is "generalized" in the sense that the relations between any two parties is presumed to be non-symmetrical (i.e., not mutual) and to involve only a unidirectional transfer of favors from one party to the other. The repayment/reciprocation of the incurred debt is not a direct return of favors at any future event, but rather a generalized return toward some other (third) party in need of a favor similar to the one formerly received. The resulting event configuration is one of "chaining" rather than "symmetry." In the social sciences, such concepts of generalized reciprocity are discussed in reference to the relationship between the generations (Kohli & Künemund, 2003; Stegbauer, 2010), but also with respect to the principles of exchange among tribal and nontribal people (e.g., gift exchange, public goods; see Mauss, 1990 [1924]; Herrmann et al., 2008; Molm, 2010; Kube et al., 2012), as well as the organizing principle of social norms and institutions in modern societies (e.g., pooling, charity, public goods; see Fehr & Fischbacher, 2004; Nowak & Sigmund, 2005; Stegbauer, 2010; Cook et al., 2013).

A notable example of a language that has conventionalized the expression of chained reciprocity in reference to social relations is the Australian language Nunggubuyu (see example [26]).

(26) NUNGGUBUYU

a. *wu:=lhama-**nyji:**-na*
3Pl=beget-Recip-Pres
'They (the old men / fathers of the clan) have children (who remain in the clan).' (Heath, 1980, T72.1.3)

b. *wu:=yaba-**nyji:**-ni-yinyung* *arawindi*
3Pl=bear-Recip-Pst-Rel many
'they who gave birth to each other (i.e., the people on the mother's side)' (Heath, 1980, T157.13.3)

In the two examples above, the reciprocal suffix *-nyji* is used in construction with semantically a-symmetric verbs roughly translatable as 'to beget' (i.e., *lhama-* '(father) to beget (child)' and *yaba-* '(mother) to beget (child)'). When marked in this way, the resulting (reciprocal) verb forms express the notion of genealogical succession within the tribal community. In the case of *lhama-* (26a), reference is made to the patrilineal line of succession including fathers and their children; in the case of *yaba-* (26b), by contrast, reference is made to the matrilineal line of descent involving mothers and their children. In both cases, the meaning expressed does not relate to any transfer of favors between the generations but is instead focused on the cyclic re-instantiation and continuation of genealogical and cultural identity within the group (Heath, 1984, p. 391). The distinction between patrilineal and matrilineal descent is known to be culturally significant in many Aboriginal tribes.

In drawing attention to different concepts of social reciprocity (direct versus generalized) and the possibility to express such notions via reciprocal constructions, linguists do *not* imply that the range of application of reciprocal constructions to different situational configurations (e.g., symmetry versus chaining) is only allowed in reference to forms of social exchange. Rather, they mean to show that most of the situational configurations prototypically expressed by reciprocal markers in different languages are also, on principle, identifiable in the different *nonlinguistic* conceptualizations of social (exchange) relations. The social exchange of favors may be immediate and bidirectional (e.g., 'giving each other food,' 'grooming one another,' 'helping each other in pursuit of the same goal'), sequential and bidirectional (e.g., 'exchanging gifts,' 'mutual hospitality,' 'exchange of favors'), or generalized and chained (e.g., 'helping others,' 'punishing general offences,' 'begetting one another'). This spectrum of complex situational configurations within which human individuals constantly interact with and relate to one another bears some obvious resemblance to some of the basic parameters (and variables) set up by linguists with respect to the reciprocal construction (Majid et al., 2011a, p. 3). We return to this point in the discussion below (section on "Reciprocal Schemas").

Reference to Non-Symmetrical Events

A third observation concerns the fact that language users sometimes employ reciprocal constructions in reference to clearly non-symmetrical, unidirectional situation types, and yet (thereby) impose a sense of social reciprocity which would otherwise not be present. Support comes from examples presented in Evans et al. (2011a) in context with the video-clip material tested on informants from twenty different languages.[5] Some of these video clips included unidirectional social scenarios in which, for instance, person A talks to person B but without receiving any direct (talking) response from the latter; or, in which a person A gives an object to person B who does not reciprocate the action. In response to such nonreciprocated clip scenarios (informants were asked to provide verbal descriptions for each of the previously witnessed clip scenarios), informants from different languages varied with respect to their employment of reciprocal constructions. For some, it was out of the question to use the reciprocal construction, even if asked specifically by the conductor of the experiment. In other cases, however, informants willingly employed the reciprocal construction of their mother tongue. The reason for such deviant application of the reciprocal marker can indeed be explained in terms of social reasoning (Rapold, 2011, pp. 69–70, on Khoekhoe; Enfield, 2011, pp. 139–40, on Lao). Enfield (2011) remarks with respect to the Tai language Lao: "the responses to video stimuli clearly show that *kan3* may unproblematically mark events of non-reciprocated giving [. . .]. Perhaps the decontextualized nature of the watch-giving scene (the videoclip is only a few seconds long) makes the event look like it is part of a game the actors are playing" (Enfield, 2011, p. 140).

Further support comes from the Australian language Ngiyambaa (Donaldson, 1980). In certain contexts, speakers rely on the reciprocal suffix *-la* not as a marker of direct reciprocity, but to indicate a general/indirect sense of social obligation and mutual dependency accompanying some unidirectional action. Donaldson (1980) only provides one example of this kind (see example [27] in the quotation below[6]), but her interpretation supports the above claim:

> On another occasion someone wanted a cigarette and began to cadge for one among the people present. One of the party explained:
>
> (27) *ŋu-dha-**la**-nha*
> give-Refl.Foc-Recip-Pres
>
> which she paraphrased, 'Someone else has to give her, she's not got any.' Literally, this is '(They) give to each other, each for herself,' a succinct expression of the co-operative principle that continues to govern the distribution of goods among communities of Aboriginal descent in western New South Wales. (Donaldson, 1980, p. 178)

Moreover, in chapter 6 (section "The Mantharta and Ngayarda Languages"), we drew attention to functional extensions of the collective-reciprocal marker into the domain of social kinship marking in some Ngayarda languages. This phenomenon was specifically analyzed in Dench (1987) with respect to Martuthunira and Panyjima. The verbal suffix which typically occurs in plural-subject constructions in reference to either collective or reciprocal actions sometimes also marks the verb in singular-subject constructions. Here, its presence indicates something like the *potential* of collective activity between the participants involved (see examples [28] and [29]).

(28) MARTUTHUNIRA

Mir.ta	*yimpala-npa-**marri**-layi*	*kartungu-u*	*mapuji-i.*
not	like.that-Inch-Coll-Fut	2Sg.Gen-Acc	mother's.father-Acc

'Don't be like that about/on your grandfather.' (Dench, 1995, p. 69)

(29) PANYJIMA

Nyinta	*wiya-**nmayi**-nha*	*marrkara-ngarli-ku-rla?*
2Sg.Nom	see-Coll-Pst	brother-Pl-Acc-foregrounding.clitic

'Did you see your brothers?' (Dench, 1987, p. 330)

Dench's (1987) hypothesis is built on the observation that the collective suffix is only employed in such singular-subject constructions if the two referents (encoded in different syntactic positions) belong to the *same alternate generation sets* (not if they belong to different generation sets). As Dench argues in some detail, belonging to the same or different generation sets has important implications in the Aboriginal culture for the kind of social interactions permitted to take place between the individuals. Only if people belong to the same alternate generations sets (e.g., "grandfather-grandchild" or "brother-brother") are they allowed to engage in all sorts of collective activities (e.g., during initiation ceremonies). This is not the case between subjects who belong to different generation sets (e.g., "father-child" or "mother - son-in-law"). Against this background, the presence of the reciprocal-collective suffix in singular-subject constructions seems to indicate an increased awareness in Aboriginal speakers for the existence of collaborative ties between the individuals. And such marking persists even in reference to situations where no actual mutual or collective interactions take place.

ARGUMENTS *AGAINST* A CONNECTION

We now turn to arguments against the claim that reciprocal constructions are dedicated means for the expression of social exchange relations among humans. Some of the relevant arguments have already been put forward by König (2005) and Majid et al. (2011a) themselves when reflecting critically upon their own findings; others have not and will therefore be added to the discussion.

The Biological-Sociological Perspective

The biological-sociological predictions with respect to the semantic encoding of reciprocity in language are not confirmed by most of the empirical language data. In part, this is also concluded by König (2005):

> The concepts of reciprocity used in biology, the social sciences and philosophy are *much narrower* than the ones delimited by formal means in natural languages. The relevant concepts are exclusively positive, social forms of interaction that centre around prototypes like 'exchange,' 'sharing,' 'the gift,' 'hospitality,' 'altruism,' and 'mutual knowledge.' Moreover, anthropologists and sociologists have underlined the importance of a temporal delay between the receiving of a gift and the act of reciprocating and thus the *sequential nature* of reciprocity. (König, 2005, p. 18; emphasis added)

The biological concept of reciprocity (reciprocal altruism, cooperation) is first and foremost based on the notion of *non-symmetrical event configurations*. It focuses on encounters between two or more individuals, who repeatedly interact with one another under variable external conditions. Classical examples of such behaviors in animals are the food exchange among South American vampire bats (Wilkinson, 1984), or the different forms of reciprocity displayed among chimpanzees in such different domains as feeding, grooming, or conflict resolution (De Waal, 1997, 2005a, 2005b; Engelmann et al., 2015). With respect to humans, anthropologists and sociologists concentrate on ritualized conventions or ad-hoc negotiations and role assignments in the context of the material/immaterial transfer of goods/favors (Mauss, 1990 [1924]; Berg et al., 1995; Tomasello, 2009; Stegbauer, 2010; Molm, 2010; Cook et al., 2013). Hence, concepts of social reciprocity center around such notions as 'exchange,' 'sharing,' 'mutual tolerance,' 'collaboration,' and so on.

A conspicuous feature identified in all of these cooperative interactions is thus the non-symmetry of relations and role assignments holding between the individuals (reciprocants) interacting in a given situation. In any single encounter between two individuals (A, B), cooperation does not consist of mutual-symmetric interactions between individual A and individual B, but of directed, non-symmetric interactions. Individual A invests in individual B without receiving any direct benefit/reciprocation from B. Hence, social exchange is prototypically characterized by some indefinite time delay between the giving and the receiving of a favor.

From the biological perspective, the reason for this immediate non-symmetry relation between cooperating individuals is due to the *maximization of individual gain and loss accounts*. By repeatedly engaging in altruistic behavior, individuals are able to maximize their own benefits. According to the theory, cooperators ultimately achieve higher levels of individual payoff than non-cooperators do. From an evolutionary perspective, the key lies in the observation that such cooperative encounters tend to come at a relatively *small cost* for investing individuals while they tend to come at an *immense gain* for receiving individuals (Trivers, 1971). Individuals who act altruistically toward others in a given situation do so at an energetically low cost for themselves, while individuals who receive such altruistic support from others experience a great increase in individual fitness. Given that the environmental conditions, under which all group members live together, are variable, and given that each group member might potentially face sudden dilemmas, the possibility to receive help from other group members in such situations can, on principle, result in a great selective advantage for *all* group members.

We now come to the consequences with regard to the encoding of reciprocity in language. If the employment of reciprocal constructions in language is truly a dedicated means for expressing social forms of exchange, we would predict (1) that the emergence and distribution of reciprocals is specifically bound to the expression of social exchange, and (2) that the relevant situations expressed in this way bear the non-symmetrical, directed event configuration so characteristic of the prototypical cooperative encounter. However, neither of these predictions is confirmed by the linguistic data.

With respect to the first prediction, König (2005) already provided the most conspicuous counterargument to a connection: "The concepts of reciprocity used in biology [and] the social sciences [. . .] are much narrower than the ones delimited by formal means in natural languages" (König, 2005, p. 18). The distributional patterns of reciprocal constructions are by no means restricted to the expression of only social forms of exchange. Instead, the range of meanings expressed is far more inclusive and, moreover, unspecific as to the social value attributed to the actions/states depicted in this way. Reciprocals are used "for all kinds of symmetric relations and interactions" (König, 2005, p. 5), ranging across such different predicational meanings as 'touch,' 'help,' 'talk,' 'give,' 'hate,' 'see,' 'ruin,' 'look at,' 'meet,' 'know,' 'envy,' 'compare,' 'resemble,' 'chase,' 'follow,' 'disperse,' 'be parallel,' 'connect,' 'mix,' 'stack on top of,' 'sit next to,' or 'be related.' Moreover, the formal and semantic restrictions of reciprocal marking patterns never run along the lines predictable from the biological-sociological perspective. For example, they are not typically restricted by distinctions like positive versus negative interaction, direct versus generalized reciprocity, social versus non-social interaction, social cooperation versus interactions without cooperation, and so on. Instead, "linguistic concepts of reciprocity completely lack any evaluative component" (König, 2005, pp. 18–19) of this kind.

With respect to the second prediction, we note that reference to non-symmetrical, directed event configurations by means of reciprocal markers is extremely rare. From the biological perspective, the *prototypical* use of reciprocal constructions should be somewhat similar to the marginal uses of reciprocal markers outlined earlier in this section (nonreciprocated events in Lao; the indication of a general sense of mutual obligation and dependency in Ngiyambaa; the expression of potential collective activity in Ngayarda languages). Perhaps the latter two examples in particular come closest to the biological-anthropological expectation of what social reciprocal marking in a language should look like. The special features of both these reciprocal marker functions are their reference to clearly directed, non-symmetrical event configurations (i.e., 'A → B,' but not 'A ← B'), and their indication, despite apparent non-symmetry, of the socially cooperative principle generally guiding such directed actions between the participants involved. Thus, according to Donaldson (1980), the use of the reciprocal suffix *-la* in Ngiyambaa indicates a distinct awareness of speakers that the described action presents a transfer of favor ('giving someone a cigarette') that is specifically *evaluated* against the cooperative principle penetrating their entire social system. This use of the reciprocal marker thereby acknowledges both, the non-symmetry of the cooperative situation as well as the general tit-for-tat pattern motivating such directed transactions in social behavior.

Earlier, the Ngiyambaa example and others were presented as an argument *for* a connection between reciprocal marking in language and the pragmatics of social exchange. However, we need to put such examples into perspective. First, they are extremely rare cross-linguistically. They are not at all representative of the typical uses of reciprocal markers in most languages, and even in those languages in which they are found pose an exception rather than confirmation of the dominant reciprocal marking pattern. And second, these marginal uses of the reciprocal marker seem to be merely optional, not grammatical, obligatory devices; it seems possible to make the same utterances without the reciprocal marker present. In such cases, the presence or absence of the reciprocal marker seems to impact only on the pragmatic implications of the utterance, but not its idiomaticity or grammaticality (well-formedness).[7] Of course, all this is not to say that language data of this kind does not carry some weight or pose interesting questions. In fact, it is quite remarkable and, from the linguistic perspective, unexpected that such forms of language use should exist. Their occurrence in small speech communities where the interpersonal relations and mutual dependencies among people are felt so strongly at all times should probably say something about the strong effect that culture may have upon language use. Nevertheless, this exceptional recruitment of reciprocal markers for the expression of cooperative behavior also demonstrates how marginal such conjectures in language use are, and how different they are from the much more typical employment of reciprocal markers as mere expressions of symmetry in whatever context (social, perceptual, physical, spatial etc.).

The Language-Typological Perspective

From the linguistic (language-typological) perspective, characterizations of reciprocal meanings as well as the distributional patterns associated with reciprocal markers do not yield any positive link to the biological-sociological characterization of social reciprocity. There is no significant correlation between cross-linguistic variation in reciprocal marking and cross-group variation in social exchange. Indeed, the presence or absence of dedicated reciprocal marking in a language seems to have nothing to say about the nature or importance of social exchange within a given society. Moreover, the so-called granularity of event description (Evans, 2007) which is typically associated with the use of reciprocal constructions tends to abstract away from the social-cooperative aspect of a situation, thereby running counter the claim that reciprocals are dedicated instruments for the expression of socially reciprocal behavior. These various considerations are all strong arguments against a connection. Next, we discuss each consideration in turn.

Reciprocal Schemas

In many, probably most languages around the world, reciprocal constructions are used in reference to a plurality of (human/animate) participants. The minimum number is two, but there may be more than two (e.g., English *They are fighting each other* 'they (two or more) are fighting each other'). Yet, in some languages, deviations from this prototype occur. For instance, in the Australian language Ngiyambaa (New South Wales), it is not possible to use the reciprocal verbal suffix *-la* unless all participants are encoded by a *single* noun phrase in subject position. Hence, it is not possible to produce sentences like 'The dingoes and the dogs are biting-Recip.' Instead, to encode the reciprocal relation, two separate sentences must be produced: 'The dingoes are biting the dogs. The dogs are biting the dingoes' (Donaldson, 1980, p. 168). In other languages, users show a preference to employ reciprocals in reference to dyadic interaction, that is, actions involving just two participants. "An example is the Taiwanese language Paiwan, which forms dual and plural reciprocal meanings by a different reduplication pattern: *pacun* 'look,' *ma-pa-pacun* '(two) look at each other,' *ma-pa-pacu-pacun* '(more than two) look at one another' (Zeitoun 2002)" (Evans et al., 2011b, p. 9). Another example is the Australian language Mawng where speakers, exposed to the video-clip material by Evans et al. (2004), more readily employed the reciprocal construction in reference to actions involving two people as opposed to more than two (Singer, 2011, p. 247).

Another parameter is the degree of differentiation (elaboration) between reciprocating events and their temporal composition. Overall, in comparison to other construction types, reciprocals are characterized by a relatively low degree of elaboration of events (Kemmer, 1993, p. 208; see also chapter 5). Thus, the spelling out of the reciprocal relation in a bi-clausal construction (e.g., English *Mary looked at Peter and Peter looked at Mary*) expresses a larger degree of elaboration of event description than a monoclausal construction with some dedicated reciprocal marker (i.e., English *Mary and Peter looked at each other*). However, there is still room for different conceptualizations of such "unitary action." In particular, the differentiation between simultaneous as opposed to sequential action plays a significant role in many languages (Evans et al., 2011b, p. 9). English already exemplifies this distinction. Language users employ lexical reciprocals (e.g., *They kissed*) differently from grammatical reciprocals (*They kissed each other*) depending on the simultaneity and immediacy of action: the lexical reciprocal implies a higher degree of simultaneity ('two or more people kissing one another simultaneously [on the mouth]'); the grammatical reciprocal implies more sequentiality and/or individuality of action ('two or more people kissing each other alternately [on the cheek]'). Similarly, in chapter 7 (section "Semantic Context Factors"), we gave another example from the Aslian language Mah Meri (Malay Peninsula). In response to the video-clip study (Evans et al., 2004), informants of the language showed a marked tendency to use the distributive-reciprocal construction even with naturally reciprocal verbs in reference to slightly a-typical mutual event configurations (Kruspe, 2011, p. 159). By contrast, in cases where the witnessed scenarios were more ordinary and inferable from common knowledge, naturally reciprocal verbs were used as lexical reciprocals. A further example is the Australian language Warrungu, a near neighbor of Yidiny (see Map 4.1, p. 68). Here, the reciprocal verbal marker *-wa* is sometimes accompanied by the (optional) reciprocal adverb *paya-paya* 'in one's turn, in return, in turn,' which in such cases conveys an alternate reading, as opposed to a simultaneous one (e.g., 'we (two) gave [a possum] to each other in turn'; Tsunoda, 2007, pp. 1419–20).

Reciprocal markers also show different sensitivities toward dynamic as opposed to stative events, as well as the degree of mutuality or symmetry witnessed in a situation. For example, in the Aslian language Jahai, neither of the two reciprocal constructions is used by speakers in combination with stative verbs (Burenhult, 2011, p. 166). Likewise, in the closely related Mah Meri, "clips depicting symmetrical physical states, e.g., 'be.next.to,' 'lean,' were impossible to describe with the reciprocal construction because intransitive and stative verbs cannot enter into reciprocal constructions" (Majid et al., 2011a, p. 40; for the primary data, see Kruspe, 2011). Mah Meri is also special because its speakers show more awareness toward the degree of symmetry and/or sameness depicted in a real-world scenario. Thus, while conducting the video-clip experiment, Krupse encountered some difficulties to record event descriptions by means of reciprocal constructions at all:

> Lexical items in Mah Meri often exhibit detailed semantics, with no corresponding generic verb. An example of this was encountered with the "hitting" scenes, where each participant "hit" in a different manner, either with a different degree of force, or making contact with a different part of the body, and so forth. In response, consultants produced blow-by-blow multiclausal descriptions of the event (as with #42, where three pairs hit each other). Once this problem was identified, responses were prompted on the basis of the actions being constant, generally resulting in a double distributive construction as shown with "#5, modified" in Example (12) above. (Kruspe, 2011, pp. 157–58)

In many other languages, reciprocals are far less sensitive to the degree of symmetry entailed by a situation. They are more focused on the plurality of actions and the fact that various directed actions take place at more or less the same time. In particular, this is a characteristic associated with collective-reciprocal marking patterns (e.g., the Ngayarda language Martuthunira; Dench, 1995; see also chapter 6 and the discussion in Lichtenberk, 1985, 1999). Moreover, in many cases, the degree of mutuality accepted and/or required under the use of a reciprocal construction depends on the specific type of event (e.g., hitting, seeing, meeting, exchanging objects). Thus, the empirical studies in Evans et al. (2011a) provide many examples where speakers of a language accept various a-symmetries of event configurations and still use the reciprocal construction. Especially in those instances (video-clip scenarios) where actions or states deviated from everyday experience, language users showed extreme freedom and variability in relying on linguistic constructions to communicate what they had seen. In some cases, this meant that speakers would rely on reciprocal constructions to implicate special arrangements of plurality, simultaneity, strong symmetry and/or sequentiality of action. In others, it meant that speakers would rely on entirely different (nonreciprocal) constructional means in order to spell out complex event configurations.

To summarize, the meaning of reciprocals is based on various semantic parameters like number, animacy, temporality, and symmetry/mutuality (Evans et al., 2011b; Majid et al., 2011a). Although the typological literature does not tend to dwell on this point, it is worthwhile to note that these various parameters describe extremely coarse-grained ("schematic") conceptualizations of real-world experience. It is not possible to align these *reciprocal schemas* with the prototypical biological-social reciprocal event configuration, nor can the latter motivate the patterns of variation that recur across languages. Ideally, from the linguistic perspective, we would expect to find the same amount of cross-group variation in social exchange as we find in language use. We should be able to distinguish, say, between societies engaging in *dyadic forms of social exchange* (only involving two participants) as opposed to non-dyadic forms of exchange (involving two or more participants); we should expect to distinguish between societies engaging in *dynamic forms of exchange* as opposed to others engaging also in non-dynamic (static) forms of social reciprocity; or we should be able to distinguish between societies engaging only in *strictly simultaneous and/or symmetric forms of exchange* as opposed to others also allowing sequential and/or slightly non-symmetrical forms of exchange; and so on.

It is likely that neither linguists nor sociologists will be interested to test such hypotheses empirically. It seems too obvious that forms of social exchange in different societies are not built on such abstract, *nonevaluative* parameter settings. However, at least some little evidence would be required as positive confirmation of the claim that reciprocal constructions are dedicated means for the expression of social exchange relations. If such evidence cannot be gathered, there is also no basis for the hypothesis (strong interpretation) that reciprocals are "grammars of exchange" (Majid et al., 2011a) incarnate.

Granularity of Event Description

Another, not wholly unrelated, feature of reciprocal constructions—which helps us differentiate between the social and linguistic concept of reciprocity—concerns their *granularity of event description* (Evans, 2007). As is generally characteristic of language use, speakers are free to choose between quite different levels of granularity of description, ranging from fine-grained to general (Kemmer, 1993, p. 208; Van Valin & LaPolla, 1997, pp. 82–102). In reciprocal constructions, the chosen level of granularity tends to be *general rather than fine-grained*. Compare, for instance, the more fine-grained description by means of a complex (bi-clausal) sentence in example (30a) with the more general description by means of a simple reciprocal construction in example (30b)[8]:

(30) ENGLISH
   a. *John gave Mary an apple, and Mary gave John an orange.*
   b. *John and Mary gave* ***each other*** *food/fruit/a piece of fruit.*

If a speaker chooses to be specific about the theme of transfer (e.g., type of fruit), they are compelled to spell out the relations between A and B in a succession of separate clauses (example [30a]). If, on the other hand, for some reason they do not mind the specificity of the theme, they may as well choose a more general level of description (by denoting a more generic "food item" category). The latter choice allows them to employ the symmetric *each other*-construction (example [30b]). Thus, "what is asymmetric at one level of description can be symmetric at another level of description" (Evans, 2007, p. 21). It all depends on the intentionality of the (linguistic) action.

This flexibility of speakers to choose different levels of granularity in reference to (simple or complex) real-world situations makes it possible for them to uncover symmetry not just on different semantic *levels* but also in focus on different semantic *aspects* of a situation. This latter point does not so much concern the degree of abstractness (general versus fine-grained), but rather the type of semantic action (e.g., social versus psychological versus physical) focused on in a given context. Speakers again have a choice, but this time between the description of physical or perceptual/sensory cues and the description of social attitudes or emotional states. In either case, these different intentional frames rely on rich pragmatic inferences regarding physical causality, social intelligence, and/or emotional states. This makes it possible for them to express the symmetric situation under either the social or physical aspect, even in cases where the description of other outward appearances would not allow such symmetry (see examples [31] and [32]).

(31) ENGLISH
  a. *John carries a stack of books to the drawer, and Mary opens the door in front of him.*
  b. *John and Mary* ***help each other*** *(put away a stack of books in the drawer).*

(32) ENGLISH
  a. *Jim looks at Bill, but Bill ignores Jim and looks in another direction.*
  b. *Jim and Bill* ***go past each other***.

Both depicted scenarios ([31] and [32], respectively) involve a form of social reciprocity (cooperation, antagonism). If focusing only on the social aspect of the situation, we can use the reciprocal construction in (31) (see example [31b]) but not in (32) (see example [32b], which merely describes a physical event). However, if taking nonsocial (physical and/or temporal) aspects into consideration, we cannot easily use the reciprocal construction in (31) (where the b-sentence only captures the social aspect), but we can in (32) (see example [32b]). In cases where mutual symmetry is not an option either in the social or the physical domain, spelling out the relations unidirectionally becomes the default strategy.

Finally, we note that the employment of the reciprocal has the effect of completely disregarding those aspects of a situation that, from the biological perspective, make it socially cooperative in the first place. In the above examples, the socially (non-)reciprocal character of each scenario is only made fully explicit in the nonreciprocal sentences (examples [31a], [32a]), while it remains entirely obscure in the reciprocal sentences (examples [31b], [32b]). The reason for this "abstraction" in the case of the latter was made clear earlier: the different participant roles and relations between participants are blended together under the aspect of "sameness" or "symmetry," so that only those aspects of the situation are focused that are applicable on either side. As a result, all those aspects of a situation that may characterize it as proper (non-)cooperative behavior are lost to the verbal description. Hence, in the two nonreciprocal sentences above we get a fair idea of the kind of social interaction referred to—in terms of the complementarity of roles, directedness, initiative, and response—while in the two reciprocal sentences all this evaluative meaning of the scenario gets lost. It may sure be inferred, but it is not encoded by the construction itself. Hence, the "communicative function" of the reciprocal runs exactly opposite to our initial claim that such constructions are instrumental for the expression of social exchange relations between humans.

This analysis is corroborated by natural language data from Australian Aboriginal languages. As we mentioned in chapter 4, the reciprocal data set collected for this study was partly based on spoken narratives (e.g., storytelling, ethnographic narratives), which were published by grammar writers as transcribed, glossed, and translated texts. In searching through such "spoken texts," it was possible to observe more closely the occurrences of reciprocals and the specific discourse contexts in which they were used. In this context, it was a conspicuous observation that the use of reciprocals often occurred either in introduction to or in conclusion of a depicted complex scenario (e.g., a talking scene, a fighting scene, a sharing scene), but was then either preceded or followed by multiclausal explanatory passages, which alone served as a proper clarification of the social content expressed. Sometimes, the grammatical reciprocal was also replaced by lexical items for the expression of symmetry. As an example, consider the following text passage, which explains 'wife-exchange' in Nunggubuyu society:

(33) NUNGGUBUYU

Ngar-uba-gi-yung / ngara-ngari-mar̲a-yung / ngara-ngari-mar̲a-yu-maji: /
that (Sg) (woman's) daughter if her daughter

ngara-ngari-lhari-yung ngara-ngari-lhari-yu-maji: / ngar-uba-gi-yung /
his sister if his sister that (FSg)

ambiringa=yaba-ng ngara-mani-nyung an-uba-gu / mana-da:n-jinyung /
they will bear her woman there of belly

ba-gi-yung / ba-gi-yung ni-ga:-'yung nu:-'ba-gi-yung / ni-ga-maji:
he is there he is there as for him that (MSg) if he

ambuni:='-ny an-uwa-ga:-'la / ambuni:='ny-maji: / ni-ga:-'yung
they give him from there if they give him as for him

ngari-mara-yung ana:-'-ji:-'li ana-wungali wu:-ngunyju=ngunyju
sister's daughter from here exchange they are equivalent

ambani:='-ny / ni-ga ambuni:='ny ni-ga:-'yung
he will give them he they will give him as for him

ambani:='-ny ana-wungali.
he will give them exchange

'Suppose that someone who is the true sister ('of the belly') of a man bears a daughter (the man's SiDa). Then if some people give that man (a wife), he will give them his new sister's daughter (as a wife for one of them). There will be an equal exchange, they will be even.' (Heath, 1980, T78, pp. 342–43)

The passage illustrates nicely how a formally dedicated marker of reciprocity, like the verbal suffix *-nyji* in Nunggubuyu (see chapters 5 and 6), is by no means the primary linguistic choice of a native speaker when making "social exchange" the actual topic of conversation: using a monoclausal reciprocal construction involving verbal *-nyji* would, in the above passage, simply not help clarifying anything about the actual procedure of 'wife-exchange.' Instead, the linguistic tools employed in the above passage are, first, a concatenation of simple sentences, each of which spells out only some part of the entire exchange procedure, and, second, the use of specific lexical items like *ana-wugali* 'exchange' and the reduplicated form *ngunyju-ngunyju* 'equal, reciprocal,' whose lexical meanings help to summarize in more generic terms the giving and the receiving in this form of transfer.

To conclude, the use of the reciprocal construction has the communicative function to express complex situations with a focus on symmetric and reciprocally directed relations between two or more participants. Speakers can establish such reference on different levels of granularity (general versus fine-grained) and in reference to different semantic event types (social, physical, psychological). In particular, the communicative value of reciprocals resides in its *concise* reference to complex scenarios and its easy clarification of the basic, schematic event configuration underlying them. Based on this communicative function, which is typical of most reciprocals cross-linguistically, the reciprocal construction does *not*, even in reference to social interactions among humans, pose an "instrumentalization" for the expression of social exchange relations. Instead, we observe that especially those aspects of a situation that, from the biological-sociological perspective, specifically define the reciprocally altruistic relationship between the participants, remain unexpressed. The only way to recover them linguistically is by way of non-dedicated, nonreciprocal, directed formulations, such as multiclausal event descriptions. The employment of reciprocal markers as "instruments" for the expression of social exchange must be considered extremely rare and exceptional. As was discussed earlier, the marginal use of the reciprocal suffix *-la* in Ngyiambaa (example [27]) in reference to directed, but socially cooperative interactions comes closest to something like a true linguistic "social reciprocity"-marker. However, despite such exceptions, the symmetric event description that disregards all aspects relevant to an understanding of reciprocally altruistic behavior, poses the semantic prototype of linguistic reciprocity. The functional versatility hereby displayed by the reciprocal marker, its broad range of application far beyond the core area of socially reciprocal behavior, underscores its psychological status as "communicative tool."

The Analogy to Prelinguistic Tool Use

Finally, the analogy to prelinguistic tool use makes it implausible to assume that grammatical forms of language use are linked to psychological dispositions for specific types of behavior. The acquisition and subsequent transfer of tools is not linked to or dependent on specific psychological attitudes or emotions in the acting individual. It is true that being hungry, curious, or playful are important intrinsic motivations for individuals to become interested in solving a problem in the first place. But *desiring* something and *knowing* how to solve a problem in serving that interest are two entirely different cognitive processes (Köhler, 1921). The specific target area is no more than an incentive for the individual to look for an effective tool with which it can achieve its goal. The tool knowledge itself remains independent of the psychological value attributed to the target area and can therefore be transferred to all manner of other target areas in which the same type of problem solving is required. Thus, Köhler (1921) observed that his chimpanzees would freely transfer the use of sticks in the "out-of-reach" function to different target areas that were not at all identical to the type of target encountered in the initial learning context: once it had been understood in a "hunger-driven" context that the stick would be useful for reaching an out-of-reach food item (e.g., a banana), it would subsequently not just be applicable to all manner of other food items, but it would just as quickly spread to entirely different behavioral contexts, where hunger would not be the driving force at all. The use of sticks would also resurface in "play-driven" contexts (e.g., reaching for inanimate objects), in "curiosity-driven" contexts (e.g., examining animals), or in "malice-driven" contexts (e.g., teasing chickens).[9]

Similarly, in the case of reciprocal constructions, the social aspect so frequently depicted by these constructions in reference to a situation seems to reflect no more than a common interest or *desire* in humans to talk about social affairs. We therefore cannot conclude that the use of reciprocal constructions is specific to the expression of social exchange relations. If we did, this would be the rough equivalent to claiming that tool use in chimpanzees is specifically bound to reaching out-of-reach food items like bananas or hard-shelled fruits. For example, what should be the core target area of stick use in the wild? Is it dipping for ants, digging for ants, fishing for the contents of beehives, disabling bees coming out of the beehive, emptying half-cracked nutshells or scooping algae? The answer is that there does not exist any specific core target area to which stick use can be specifically linked. Subjects spontaneously invent, and subsequently conventionalize, such stick uses in application to whatever obstacles they encounter in their day-to-day business. It is simply not possible to formulate any core context in which stick use will always be practiced within a given ape community (e.g., see Whiten et al., 1999).[10]

Functionalist linguists claiming a connection between linguistic and social reciprocity might object that their claim regarding a semantic core concept of linguistic reciprocity does *not* exclude cross-linguistic cultural variation (Majid et al., 2011a)—and that therefore, by analogy, the existence of cross-group variation in stick uses, too, does not prove the absence of a "semantic core" area, namely edible targets in general. However, if that should be the claim, we can no longer put forward any empirical data that could disprove the claim of the existence of a semantic core. Whenever we find cross-linguistic differences, the argument of cultural variation will be invoked.[11] The only possibility to solve this dilemma empirically is to take the data "at face value" (Evans & Levinson, 2009, p. 432). If we do, we are left with the conclusion that it is not possible to speak of a semantic core of linguistic reciprocity. It is merely possible to speak of general areas of interest in application to which many languages employ dedicated reciprocal constructions. In the face of the empirical language data presented here, it is fair to generalize that by far most uses of reciprocal markers are with verbs (or predicates) denoting social rather than nonsocial events. Thus, in many contexts, speakers seem more *interested* in the social aspects of a situation than, say, physical relations or mere visible outward appearances (Evans et al., 2011b, p. 2).[12]

This finding renders the fact less surprising that some languages completely lack the employment of reciprocal constructions even though their speakers are well known to engage extensively in forms of social exchange (see chapter 5, section "Ku Waru and Other Languages"; see also Majid et al., 2011a, pp. 11–12). The absence of dedicated reciprocal marking in a language like Kilivila or Ku Waru oratory should not be considered surprising, but simply not indicative of the elaborateness or centrality of social exchange within the respective society. Conversely, the presence of one or several dedicated means for expressing symmetrical relations in a language should likewise not be taken as a measure for the relative importance of social reciprocity in a given society. The reason is that linguistic means expressing mutual/symmetrical relationships are not targeted at the expression of social emotions or dispositions of behavior. They are aimed at the intentional mindsets of other individuals and are communicative "tools" hinting at broad schematic arrangements of real-world experience. Their interpretations are based on common grounds of experience. Such experience is naturally enmeshed with complex social emotions, and reference to reciprocal configurations may often involve pragmatic inferencing on social reciprocity. But, as we have seen, so may nonreciprocal construction types. In other words, such meaning inferences are inferred through context; they are not encoded directly by the linguistic expressions. The latter only sketch schematic arrangements and are thereby identifiable as real "tools of communication."

NOTES

1. In Kuhle (2007, unpublished manuscript), these various concepts of non-linguistic reciprocity were explored in detail and set opposite the linguistic (language-typological) concept of reciprocity. See also the discussion below.

2. Among the languages considered here, the Papuan language Kâte and the Oceanic language Fijian, too, possess such dedicated formal means for the expression of social kinship relations (Pilhofer, 1969 [1933]; Dixon, 1988).

3. For details on notational conventions with regard to language examples, see chapter 4, endnote 1.

4. In Everett's terminology, natural languages (and their components) are also "cultural tools" (Everett, 2012, 2016, pp. 104–105); however, he does not argue in favor of a significant link to prelinguistic tool cultures in great apes.

5. The video clips (a total of 64) consisted of play-acted scenarios with two or more human individuals who would interact with or relate to one another socially, physically, and/or spatially according to various parameter settings. The relevant parameters included number of participants, configuration, symmetry, temporal organization and event type. The stimulus set and procedure of testing are explained in Majid et al. (2011b). The stimulus set included scenarios like 'mutual talking between two people,' 'mutual talking between more than two people,' 'mutual giving between pairs of people,' 'unidirectional chasing between two people,' and many more.

6. The example number "(27)" follows the present work's chronology of language data (examples), not the one in Donaldson (1980).

7. Note that this statement is not meant as a "generative" argument, although perhaps it is, after all, a difficult point to make. Even functionalist linguists (must) assume that some constructions are considered idiomatic and well formed, while others are considered odd or even wrong by native speakers. Thus, linguists probe especially for those linguistic elements and their combinatorics that present the most basic, minimum "design features" that characterize the formation of simple or complex utterances. It is against this general methodological background that I venture to distinguish here between "optional" (pragmatic) and "obligatory" (grammatical) uses of reciprocal markers. However, I concede that it is still risky to phrase it in this way. It is risky because it tends to place the analysis of linguistic constructions in too abstract terms and thus remote from the actual *contexts* in which linguistic means are naturally used and judged by their users. As works like Rumsey (1990, 2010) show, native speakers of many non-European languages do not know the difference between propositional meaning and pragmatic meaning and thus their notion of grammaticality is also much more extended/inclusive than the one established in mainstream linguistic analyses.

8. The examples are taken from the discussion in Evans (2007, p. 20).

9. One memorable passage in Köhler (1921), illustrating the irrelevance of the specific target area in pursuit of which a given tool function is first acquired, is the one where Sultan has just learnt—through a moment of insight—how to insert two sticks into one another in order to reach some food items. In this context, Köhler reports that as he returned to the testing chamber, from which he (Köhler) had been absent for only a few minutes, he found Sultan enthusiastically fishing for all sorts of objects (the food items were no longer the sole object of interest), with apparently no other purpose than simply to reach for *all things* out of reach (see Köhler, 1921, pp. 91–92).

10. In chapter 3, the same finding was shown for nut-cracking behavior. Chimpanzees at Bossou use only stones to crack Elaeis nuts; chimpanzees at Taï crack open five different nuts species including Panda and Coula nuts, but they avoid Elaeis nuts; and chimpanzees at Yealé crack both Elaeis and Coula nuts, but not Panda nuts. Which of these nut species, then, present the core target area?

11. In the generative theory, the argument is typically not cultural variation, but some endogenously determined switchboard responsible for different parameter settings.

12. This linguistic observation is similar to the observation with regard to prelinguistic tool use that in many chimpanzee communities, sticks are indeed used in application to ant or termite mounds, or that the use of stones or wooden clubs, if at all present, is most likely administered to hard-shelled fruits. In other words, just as linguists have a pretty good idea of where to find dedicated means for the expression of symmetrical relations in a language, so primatologists have a pretty good idea of where to find certain types of tool practice. However, time and again, such expectations, either by linguists or primatologists, are also disappointed by individual (speech/ape) communities. Dedicated means are not found in areas where they were expected but are found in areas where they were not initially expected. Of course, both functional linguists and field primatologists are aware of this fact and, therefore, tend to promote ever more inclusive searching methods for uncovering these true ranges of cultural diversity (e.g., see Evans, 2007, with respect to reciprocals).

Chapter 9

Conclusions

In the final sections of this book, we discuss what the tool hypothesis contributes to the larger debate on language evolution, language diversity, language use, and the psychology of language. We proceed from the prelinguistic (nonhuman) to the linguistic (human) domain and then illustrate the hypothesis' relevance in cross-linguistic typological research. We conclude with the identification and positioning of the tool approach's most ambitious claims in relation to other current theories of language learning and development.

TOOL CONCEPTS AS THE EVOLUTIONARY PRECURSOR OF LANGUAGE

The notion of *tool concepts* ("tool intelligence") offers an important insight into the higher cognitive abilities of animals and humans. It allows one to capture general intelligence at its highest structural and functional level of mental representation and active performance—and still prior to the existence of human language (see Figure 9.1).

Tool concepts describe a level of information recombination (restructuring, reformulation) at which individuals begin to mentally "construct" complex schemes of problem assessment on the basis of which to "manipulate" their environment (Mayer, 1995; Call, 2013; Boesch, 2013; Menzel, 2013, pp. 644–45). In his seminal work on the mentality of great apes, Köhler (1921) took great care to design problems that would prompt subjects to invent instrumental means by which they were able to overcome some obstacle. This, Köhler noted, was sometimes difficult to achieve as individuals would try to physically maneuver around such obstacles wherever this was possible (Köhler, 1957, pp. 17–18). As indeed most research in the tradition of the gestalt-psychological literature maintains (Mayer,

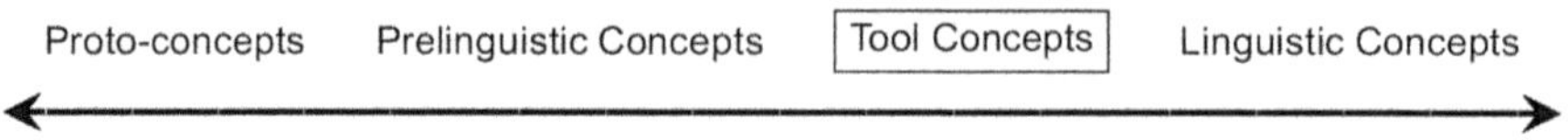

Figure 9.1 Tool Concepts in the Continuum from Prelinguistic to Linguistic Cognition. The continuum is based on the Hurford's (2007) scale of "proto-concepts," "pre-linguistic concepts," and "linguistic concepts," which was introduced towards the end of chapter 2 (sections "The General Intelligence Hypothesis" and "Finding the Missing Link"). Proto-concepts relate to instinctive, reflexive behavior, while prelinguistic concepts relate to behaviors that begin to be based on processes of generalization and flexible transfer that are considered to provide a significant foundation of intelligence and, ultimately, human intentionality. While Hurford (2007, 2012) still draws a sharp line between prelinguistic and linguistic concepts, the present study bridged this gap by introducing the additional notion of *tool concepts* ("tool intelligence"). Linguistic concepts are based on this same "tool intelligence," which in humans (and hominins, more generally) has become extended into the domain of intentional communication.

1995), there exists a considerable difference in cognitive sophistication between learning to solve problems by circumvention (physical outmaneuvering) and learning to solve them more directly, through learning by insight (see chapter 3). The mental representations underlying manipulative activities are abstract and structural not only in the sense that they rely on complex integration and recall of real-world experience. They also involve the capacity to *deconstruct* these complex experiences and to *reorganize* them *schematically* in relation to highly individual, context-dependent goal orientations. This "global mental workspace" allows the modelling of manipulations of the environment and their potential outcomes (Gärdenfors, 2004; Call, 2013; Dehaene, 2014). In gestalt-psychological terms, such "creative problem solving entails a search for a solution method that meets the requirement of a problem rather than a specific solution" (Mayer, 1995, p. 9). Under such increasingly schematic, problem-oriented and hypothetical negotiations of real-world experience, inert objects incidentally become reanalyzed as instrumental means that can serve different tool functions. Thus, a stick becomes conceptualized as a "fishing pole" or "lever"; a stone becomes a "hammer" or a "weight"; a rope becomes a "pendulum," and so on.

The behavioral evidence of such tool concepts comes from the different pattern of generalization or transfer in comparison to prelinguistic concepts based on more primary perceptions of the environment. Thus, Köhler (1921) noted the conspicuous difference between subjects who had properly learned the tool function of a stick or wooden box as opposed to those who didn't. Those who had achieved a firm structural grasp of a particular tool function were subsequently able to reapply its use very flexibly and across a wide array of situations not easily resembling the original context of acquisition (also overcoming distances of time and space). Other subjects, who failed to comprehend the tool function of an object, continued to struggle for solutions from context to context. The resulting patterns of generalization were thus

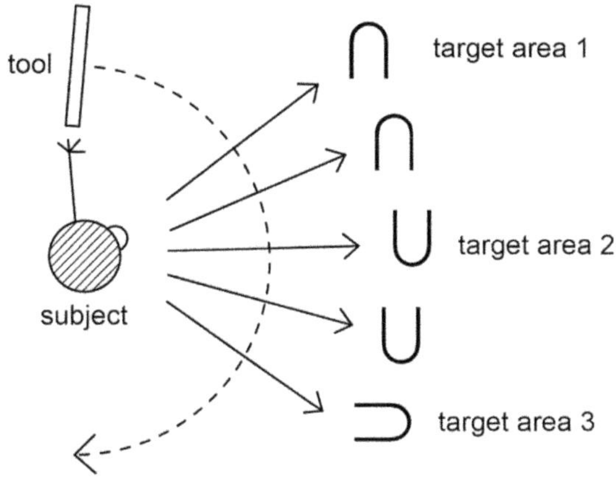

Figure 9.2 Patterns of Transfer in Tool Use. The different "arch" symbols indicate same or different problem settings that may be mastered with the same tool (target area 1, 2, 3, respectively). The pattern of transfer (dotted line with arrow at the end) indicates that tools are generalizable across wide arrays of target areas. Their functionality is problem-oriented and thereby less directed at the specific target areas ("themes") to which they are applied. Such patterns of transfer across various functional domains are highly reminiscent of the application patterns characteristic of linguistic constructions. Just as in the case of tools, their productivities are high, yet also context-bound and thus language-specific and variable. *Source*: Author created.

conspicuously broad, problem-oriented, displaced, and not confined to only specific target areas (see Figure 9.2).

Hence, tool concepts carve out new categories of functional behavior that do not (easily) align with more primary perceptions and not-tool-based interactions with the environment. The same means are applied to different ends; and the same ends are reached by different means—a phenomenon known as *means–ends dissociation* (Call, 2013, p. 10). Tool categories thus describe conceptual knowledge of the environment *sui generis*.

The ultimate (historical) effects of such tool abilities in prelinguistic animals are today studied in nonhuman primates living in their natural habitat (Whiten et al., 1999, 2001; Van Schaik & Knott, 2001; Van Schaik et al., 2003; Whiten, 2005; Boesch, 2012, 2013; Whiten, 2017). Nonhuman hominids possess a natural propensity to explore and invent such technical skills even without human scaffolding. Moreover, individuals living in different groups find different solutions to existing problems in comparable target areas. Individuals living in the same group conform to similar group-level behaviors, while those living in other groups conform to other group-level customs. Neither genetic nor ecological factors alone are sufficient to explain these differences (McGrew, 1998; Boesch, 2012). Experimental studies also confirm that nonhuman primates are sensitive to their social environment and readily imitate the actions of others (Whiten et al., 2004, 2009). Such faithful copying includes sophisticated technical skills (Whiten et al., 2005, 2007; Hopper et al., 2007; Vale et al., 2017).

Though primitive material cultures at best, these prelinguistic behaviors already display some of the characteristics that are essential for the evolutionary foundations of human language. First, they are based on an advanced *ability to manipulate* the environment *intentionally*. Second, they are of *historical origin* and do not evolve biologically in application to specific target areas (Whiten, 2005, 2017). Third, they have various detailed properties to do with preparation, application, and performance that are likely to accumulate and change over generations of tool users (Boesch, 2013, pp. 39–40). Different group customs indicate that different complexities of behavior and different ranges of application of similar techniques are the *cumulative* result of innovative usages and social transmissions within individual communities, not the result of single-step innovations of full-fledged behaviors (e.g., see the different application patterns of nut-cracking techniques reviewed in chapter 3, Figures 3.1–3.3).

The upshot of such findings is that even at this primitive stage of prelinguistic manipulative behavior, the observed form–function relationships of single tool uses are "historically constructed," that is, accumulated by individual context-dependent usages and their repeated transmissions within the group. As a result, the behavior is more "conventional" and "arbitrary" than is commonly granted in discussions on language evolution (Donald, 1991; Gärdenfors, 2004; Burling, 2005; Tomasello, 2008; Hurford, 2007, 2012; Everett, 2012, 2016). With respect to human linguistic behavior, Tomasello (2008) grants: "Communicative conventions are defined by two separable characteristics (Lewis, 1969). First and most critically, we all do something in the same way because that is the way everyone is doing it [. . .]: it is shared. Second, we could have done it differently if we had wanted to: it is, at least to some degree, arbitrary" (p. 219). In fact, as we have seen, the same requirements already apply to prelinguistic material cultures in nonhuman primates. Hence, arguably, tool concepts are the evolutionary precursors of linguistic concepts and provide the crucial "missing link" between prelinguistic concepts and the human-specific intelligence that underlies the acquisition and use of natural languages (see Figure 9.1).

HUMAN SYNTAX AND SOCIAL INTELLIGENCE—ENTERING THE HUMAN DOMAIN

Scholars concerned with the "origins of grammar" or "human syntax" commonly argue that language is fundamentally different from all prelinguistic behaviors in nonhuman animals, even primates and prelinguistic children. It is essentially "symbolic" and not externally constrained in similar ways as prelinguistic behaviors in the physical domain. For instance, Hurford (2012) goes a long way to show how prelinguistic actions, even the more sophisticated kind, always remain rooted in the physical interactions with the world. Language, by contrast, is not checked by such external physical constraints but only references things symbolically and in arbitrary form.[1]

> It is certainly safe to grant that some, but not all, physical activities with objects by gorillas, children and many other animals are hierarchically structured in the minds of the actors. But there is a key difference between such physical activity and the act of producing a hierarchically structured string of words. For the physical manipulations of objects, the current state and position of the objects themselves provide immediate sensory information about the stage reached so far in the overall process. [. . .] This is in marked contrast to the utterance of a spoken sentence. The words already spoken are no longer physically present to the senses, though they may be held briefly in a memory buffer. [. . .] What to say next when midway through a complex novel sentence is all determined by an internally kept memory of what has been said so far and the intended meaning; no external stimulus gives a prompt. (Hurford, 2012, pp. 503–4)

Given our own argument, this is an unwarranted estrangement and mystification of human linguistic behavior from its most likely psychological and evolutionary roots.

One the one hand, such positions overestimate the physical constraints and thus underestimate the cognitive demands of inventing even the simplest problem solutions by recruiting inert objects as tools. As we have seen above, such problem-solving is no longer closely molded by external constraints of the environment; it is based on intentional frames of mind (attitudes) and information recombination (reasoning) aimed at "general solution methods rather than context-specific solutions" (Mayer, 1995, p. 9). These solution methods involve complex object manipulations, not merely circumvention (physical outmaneuvering). "The chimpanzee is not simply provided for life with any special disposition which will help him to attain objects placed high up, by heaping up any building material, and yet he can accomplish this much by his own efforts, when circumstances require it, and when the material is available" (Köhler, 1957, p. 122).

On the other hand, such positions also overestimate the allegedly "symbolic" character of linguistic means and underestimate the continuous interactional-perceptual cues and constraints working on language users in the process of communication. To acknowledge that language use serves the function of communication means to acknowledge an important *external* domain that imposes natural constraints on the "construction" and "conventionalization" of linguistic behavior under variable conditions of action.

The mental states, attitudes, working memory capacities of individuals, and degrees of common grounds between them, all place immediate constraints on the processes of communication, and wherever this fails, linguistic means must be checked and renegotiated in order to achieve more desirable outcomes.[2] This ongoing negotiation with the social environment is no less *solid* than the contemplation and negotiation of tool uses in the prelinguistic domain. Material tool uses avoid physical outmaneuvering and offer problem-oriented manipulations of the physical environment. Linguistic constructions, in turn, avoid *social outmaneuvering* and offer problem-oriented manipulations of the social environment. "Linguistic acts are social acts that one person intentionally directs at another (and highlights that he is doing this) in order to direct her attention and imagination in particular ways so that she will do, know, or feel what he wants her to" (Tomasello, 2008, p. 343). Or in other words, "language constitutes a toolkit that effectively supports an individual in influencing the state and the behaviour of others" (Kolodny & Edelman, 2018, p. 3) in relation to one's own. Thus, linguistic activities are manipulative behaviors and require an intelligence that allows the acquisition and usage of verbal means under variable and displaced conditions of action. Our hypothesis states that the evolutionary roots of this human intelligence lie in the tool intelligence of prelinguistic animals.

In view of current linguistic theorizing, we may further justify this hypothesis. In particular, we may relate its claims to the *social intelligence hypothesis* that has gained considerable momentum in the discussion on language evolution over the past decades.

The social intelligence hypothesis makes two central claims with respect to the human-specific ability to acquire language (Tomasello, 1999; Tomasello et al., 2005; Moll & Tomasello, 2007; Tomasello, 2008, 2009; Tennie et al., 2009). First, already quite early during ontogenetic development, human children display a great interest in the facial expressions, focus of attention, and actions of other human beings. They possess an intrinsic motivation to engage with the feelings, intentions, and actions of others, which is why, from an early age onwards, they altruistically help, inform, share, and cooperate with others. Second, humans alone possess the higher cognitive ability to understand and reflect on the feelings and intentions of others (theory of mind) in the context of both competitive and cooperative interactions. They understand that others are and think like them and they can think recursively about each other's thoughts ("I know that you know that I know etc."), which enables them to mentally anticipate (and manipulate) each other's intentions and actions by several turns. Both intrinsic motivation for cooperation and the ability to understand others' states of mind enable humans to engage in actions of *shared intentionality*, which is the crucial psychological infrastructure on which human language is built.

> Shared intentionality involves, most basically, the ability to create with others joint intentions and joint commitments in cooperative endeavours. These joint intentions and commitments are structured by processes of joint attention and mutual knowledge, all underlain by the cooperative motives to help and share with others. [. . .] As they grow, human children are equipped to participate in this cooperative groupthink through a special kind of cultural intelligence, comprising species-unique social-cognitive skills and motivations for collaboration, communication, social learning, and other forms of shared intentionality. (Tomasello, 2009, pp. xiii–xiv, p. xvi)

As I discuss elsewhere (Kuhle, 2018a), this line of reasoning has a rather dramatic effect on the overall language debate. It redefines the notion of general intelligence in terms of social intelligence and concludes that prelinguistic animals do not yet possess general learning abilities that could serve as evolutionary precursors to the human ability to engage in intentional communication, including the acquisition of linguistic behavior. In this view, the most advanced prelinguistic skills by nonhuman animals are restricted to individualistic, physical interactions with the environment. This "individual intentionality" (Tomasello, 2008) does not carry over into the social domain. Individuals are not able to learn from one another and to socially transmit behavioral variants within the group. Social group customs are therefore not the result of cultural transmission and behaviors are not subject to historical, cumulative development. In short, prelinguistic animals (great apes) do not yet provide a useful model of learning and cultural transmission applicable to human behavior and human language in particular.

> Human cultural traditions can and do accumulate modifications over time, thus producing improbable design that survives over multiple generations. Our explanation of this phenomenon for many years has focused on the fact that human social learners focus to a much greater degree than other non-human primates on the actual actions performed by others (process copying), not just the results produced on the environment (product copying). But added to that we believe that uniquely human forms of cooperation make human social organization in many ways different as well, as things such as teaching and norms of conformity contribute to the cultural ratchet. (Tennie et al., 2009, p. 2413)

The tool hypothesis offers a different view. It claims that prelinguistic tool use abilities in great apes do provide such a model. In fact, it claims that the general intelligence hypothesis cannot do without the tool concept in order to establish cognitive and evolutionary continuity within the hominid lineage. As I argue in Kuhle (2018a), the social aspect may be specific to humans in the sense that humans engage in *triadic interactions* that require an extended awareness for the synchronization and manipulation of physical and social interaction at once. However, the notion of tool intelligence is still key to understanding how humans recruit linguistic items under displaced conditions and in a flexible, context-dependent manner. Furthermore, it is also able to account for the fact that humans and prelinguistic apes differ in their abilities to socially manipulate their environment. Apes also readily engage in social outmaneuvering and synchronization, as the "helping" and "collaboration" studies by Tomasello and colleagues suggest; however, they do not yet further *extend* this awareness to problem-solving by manipulative means within the domain of communication (Kuhle, 2018a, chapters 3.1 and 3.2). It is a notorious weakness of the social intelligence hypothesis to deny the tool character of linguistic means and the underlying psychology of behavior that bears true homology with the general intelligence in prelinguistic animals.

As we have shown in chapter 3, the tool analysis provides distinct *criteria of cultural variation.* The social intelligence hypothesis, by contrast, does not yield such testable criteria. The reason is that it focuses exclusively on the social domain of communication, which is undeniably unique to human linguistic behavior. Language, unlike prelinguistic tools, is not aimed at the physical world but at the intentional mental states (attitudes) of social others. Yet, as we maintain here, the underlying psychology sustaining the handling of linguistic means and their social transmission within the group may still be homologous to the psychological substrate of prelinguistic tool practices. The social intelligence hypothesis throws out the baby with the bathwater by discarding the tool concept in favor of the sociality aspect. The explanation of human languages requires the consideration of both aspects, but the notion of tool intelligence alone can explain the cognitive continuity to prelinguistic behaviors in nonhuman primates. The study of reciprocal constructions and other linguistic phenomena confirms this view.

CROSS-LINGUISTIC VARIATION AND FAMILY RESEMBLANCE

The cross-linguistic patterns of distribution regarding dedicated forms of language use are in principle the same patterns of distribution that emerge from cross-group comparison between nonlinguistic forms of tool use among great apes (see Figure 9.3).

As the evidence for *geographic variation* in prelinguistic tool use behavior is taken as the key argument for its *cultural-historical origin*, the same case can be made for cross-group variation in language use. The merit of constructing this analogy lies in the assessment of the cognitive substrate underlying both forms of cultural behavior. By setting the use of linguistic material in analogy with the use of nonlinguistic tools, we attain a useful perspective on

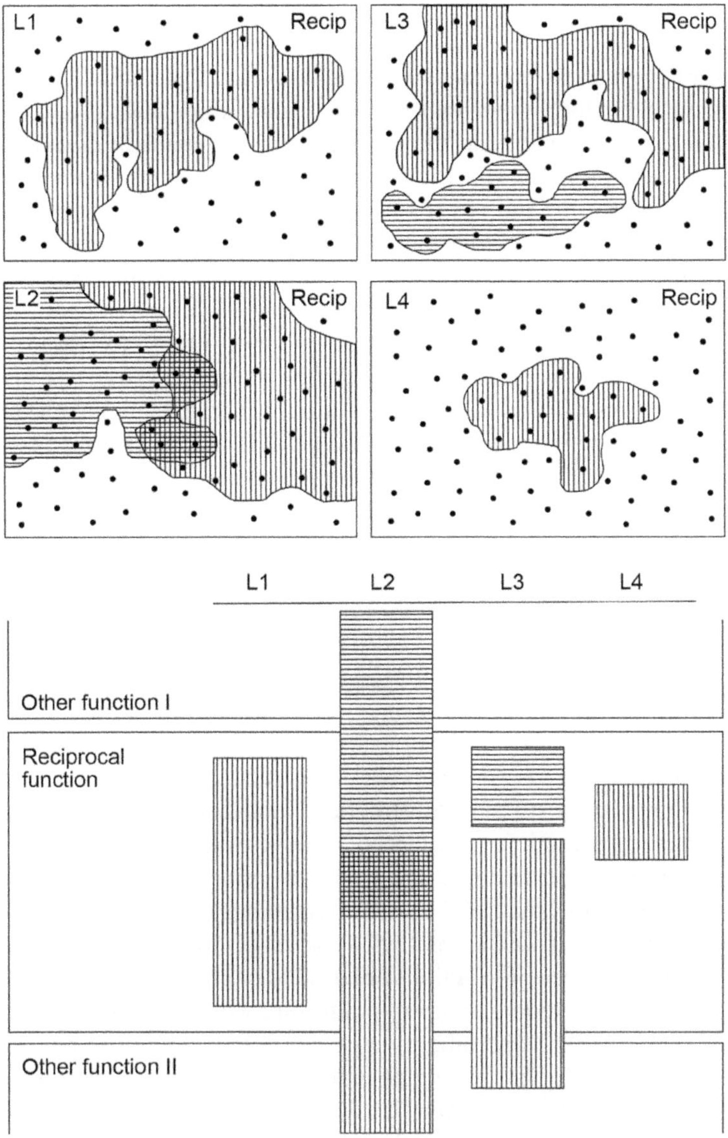

Figure 9.3 Schematic Representation of Cross-Linguistic Variation and Family Resemblance. The two (complementary) figures are based on the schematic depictions developed throughout chapters 5 and 6 with respect to the reciprocal marking patterns in individual languages. The dotted frames in figure part I (top) indicate the "possible spaces" of reciprocal marking ("Recip"). The different shadings in each frame indicate typical distributions of reciprocal markings in individual languages (L1–L4). Some of them overlap, that is, are equally productive with the lexical material (•) of a language; others are entirely in complementary distribution. Figure part II (bottom) illustrates the same marking patterns with respect to their possible functional extensions into other functional domains (i.e., nonreciprocal meanings). Such patterns of cross-linguistic variation and family resemblance are fully identical with cultural patterns of variation in prelinguistic tool uses (see Figures 3.1–3.3, end of chapter 3). *Source*: Author created.

the kind of cognitive substrate and functional value that should be attributed to linguistic behavior. In this light, specific forms of language use, such as the employment of reciprocal constructions, assume the status of *communicative tools*. As such, they are specifically aimed at overcoming communicative problems in which the individuals have to direct and influence each other's attention and ideas in reference to things under displaced conditions. To repeat Kolodny and Edelman (2018), whose argument is in broad sympathy with our own, "language constitutes a toolkit that effectively supports an individual in influencing the state and the behaviour of others" (p. 3).

The reciprocal construction as a complex construction type is one such example of a structurally dedicated, culturally variable, linguistic tool for solving a particular type of communicative problem (see chapters 4–8). The tool hypothesis is, however, in principle applicable to a much wider range of linguistic phenomena; indeed, ideally, to all. In the following, we provide some further examples of grammatical devices that have received attention in the functional-cognitive literature and that are easily accessible to the tool analysis.

Possession

The expression of possessive relationships across languages yields very similar findings as in the case of reciprocal constructions. In terms of structure, the overall range of possessive constructions includes attributive, predicative, and external constructions (McGregor, 2009; Aikhenvald, 2013). Cross-linguistically most pervasive are attributive (adnominal) constructions in which the expressions that refer to the *Possessum/Possessee* (referent who is possessed) and *Possessor* (referent who possesses) form a noun phrase (NP). Attributive possessive constructions are used to express a wide range of associative meanings, including ownership, kinship, whole-part relations (including body-part relations) and other types of associative relations between two or more referents. Typical examples include phrases like in English *John's house* (ownership), *John's father* (kinship), *John's arm* (body part), *John's hope* (general association), or *John's office* (general association). As is noted in the literature, the term *possession* is ambiguous in that "in everyday usage the term 'possession' is more or less equivalent to 'ownership' [. . .]. In traditional grammatical usage [by contrast] 'possession' and 'possessive' are construed much more broadly. [. . .] it is only a minority of what are traditionally called possessive constructions that have anything to do with property or possession" (Lyons, 1977, p. 722; cited in Aikhenvald, 2013, p. 2). Similar as with respect to reciprocal meanings, it is therefore not possible to speak of *one* cross-linguistic core function of possessives that corresponds closely with any social and/or behavioral concept of human action-perception (e.g., possession in the sense of "ownership"). Instead, as many authors agree (e.g., McGregor, 2009, p. 1; Lichtenberk, 2009, p. 262; Aikhenvald, 2013, p. 3), a whole spectrum of central or "core" functions exist that in many languages cover ownership relations, whole-part relations (including body parts) and kinship relations as well as associative relations more generally including considerations of control, use, manufacture, or location (e.g., English *John's office* which may refer to the office where John works rather than the office he owns).

However, usage-oriented factors play an important role, or as McGregor (2009, p. 1) notes, "linguistic factors" rather than "purely conceptually" defined notions of possession are most relevant in explaining the reliance on different linguistic constructions in reference to different types of possession. In this context, the distinction between *inalienable possession* and *alienable possession* has been shown to be particularly significant cross-linguistically (Barker, 1995; Chappell & McGregor, 1995). Inalienable possession refers to those relationships between referents (Possessor, Possessee) that are considered inherent to the nature of those referents; kinship terms like 'father' or 'sister' and body parts like 'arm,' 'leg,' 'head' are typical examples of such inherently relational meanings that occur in possessive constructions like 'my sister' or 'the man's head.' Alienable possession, on the other hand, denotes possessive relations that are possible but not necessarily inherent to the relationship between possessor and possessee; for instance, *John's house* refers to an association between two referents ('John–house') that is conceivable but not inherent to the relationship between those two referents ('John' is not defined by his relationship to "house" and vice versa). In other words, kinship and body-part terms inherently associate with their possessors, while many other referents (objects, persons, ideas) do not implicitly do so, but rather stand on their own and, moreover, allow various types of associative relationships to occur (*John's house* as the house he owns, the house he lives in, the house he likes, etc.).

Many languages are sensitive to this distinction and employ different formal strategies to mark each type of possession. The overall tendency of languages is to mark inalienable possession less overtly ("costly") than alienable possession. For instance, in the Arawak languages of Northern Brazil, inalienable possession is expressed by prefixes (e.g., Warekena *pi-yufana* 'your voice'), while alienable possession involves pronominal prefixes plus a special possessive suffix attached to the possessee (e.g., Warekena *pi-murupa-re* 'your canoe') (Aikhenvald, 2013, p. 8). In other languages, inalienable possession tends to be marked by synthetic means, that is, forms directly bound to either the possessor or the possessee, while alienable possession is expressed through analytic means, that is, free grammatical forms (Aikhenvald, 2013, p. 9). In all of these instances, the cross-linguistic tendency is to invest less formal marking, that is, to put less linguistic effort into the expression of inalienable than alienable possession. The most straightforward interpretation of these facts is that the costliness or complexity of such grammatical devices is directly correlated with the relative ease or effort as well as the intentions with which speakers communicate these different types of possession in ongoing speaker-hearer interactions.

However, we must qualify the above statement in one important respect. Namely, speakers of different languages do not always concur with regard to the classification of referents as either alienable or inalienable possession. Instead, the distinction remains culture-specific. In some languages, even kinship terms or body parts are marked more explicitly and thus are treated as "alienable possession." Conversely, persons or objects that are at first sight deemed "alienable" are referred to by less explicit formal marking and thus treated linguistically as "inalienable" (Aikhenvald, 2013, p. 4, pp. 12–13). The pragmatic motivations may have very different reasons. The Oceanic language Kairiru, for instance, has a large kinship system of some 18 terms. According to Wivell (1981, p. 54; cited in Lichtenberk, 2009, p. 277), only those kinship relations are marked as inalienable that carry importance within the social structure (e.g., 'father,' 'mother'); those, on the other hand, that are considered less important, receive alienable marking (e.g., 'mother's brother's wife,' 'great grandparent'). In Gapapaiwa, another Oceanic language, the situation is reversed. According to McGuckin (2002, p. 304), terms for persons who are peers or subordinates are possessed inalienably, while persons in authority over the speaker (e.g., 'spouse,' 'mother') are expressed by the alienable construction. Lichtenberk (2009, p. 278), in turn, observes with regard to the Oceanic language Toqabaqita the influence of lexical replacement. Old forms receive synthetic marking (e.g., *thaina-ku* 'my mother'), while lexical innovations, even if semantically closely related, are used in the non-suffixing construction (e.g., *maka nau* 'my father'). Hence, different pragmatic factors, ranging from cultural values to considerations of speaker-hearer relations to the formal makeup of linguistic materials influence and shape the emergence of language-specific patterns of possessive marking. Analogous observations were made with regard to reciprocals.

Finally, we may note that languages differ considerably with regard to the structural means they employ in order to mark (attributive) possessive relations in the first place. As is laid out in Aikhenvald's (2013) cross-linguistic review, possessive marking in noun phrases ranges from simple juxtaposition (e.g., I mother = 'my mother'; room I = 'my room'), marking on the possessor (e.g., as in English, where the genitive case *'s* attaches to the possessor noun), marking on the possessee (e.g., the suffix *-ni* in the Oceanic language Ambae as in *gamali-ni Robert* 'Robert's club house') to marking on both possessor and possessee (Aikhenvald, 2013, pp. 6–8). Moreover, the morphosyntactic classification of the relevant markers, if present at all (cf. juxtaposition), encompasses prefixes, suffixes, clitics as well as free pronominal forms. Some languages even lack nominal possessive marking entirely and only rely on clausal constructions to express possessive (even attributive) relationships (see Van Staden, 2009, on the Papuan language Tidore). In other words, there exists no universal constraint on how to mark possession in language and what is grammatical in one language may be ungrammatical in another.

From the Indo-European perspective, the Oceanic languages provide one of the most dramatic departures of what might be deemed conceivable ways of marking possessive relationships. These languages possess classes of *relational classifiers* (Lichtenberk, 1983b, 2009; Aikhenvald, 2013, pp. 20–24). These forms are required in possessive constructions—especially in connection with alienable possession—and are used as auxiliary means to carry the possessive marking as well as to qualify the possessive relationship. At a minimum, a classifier system consists of a relational classifier for "alimentary" possession (food, drinkables) and one for "general" possession (everything else). However, languages still vary considerably, with some having more than two or three classifiers (e.g., Micronesian languages) and others having an eroded system, where the number of classifiers has been reduced to only one (Lichtenberk 2009, pp. 268–72). Here are some examples of languages with two and more classifiers. Manam has two classifiers: the "alimentary" classifier *ʔana-* is used with both edible and drinkable possessions (e.g., *ʔulu ʔana-miŋ* breadfruit food.possession-2Pl:Poss 'your breadfruit (to eat)'; *botoli ʔana-gu* bottle food.possession-1Sg:Poss 'my bottle (drink container)'); the "general" classifier *ne-*, on the other hand, is used in all other cases (e.g., *tamoata asi ne-ø* man bushknife general.possession-3Sg:Poss 'the man's bushknife') (Lichtenberk, 1983a, pp. 291–94; Lichtenberk, 2009, p. 269). Fijian, in turn, has three relational classifiers: *ke-* "food possession," *me-* "drink possession," and *no-/ne-* "general possession" (Lichtenberk, 2009, p. 268). And Ambae has four relational classifiers: *ga-* "food possession," *me-* "drink possession," *bula-* "natural or valued object possession," and *no-* "general possession" (Hyslop, 2001, pp. 177–80; Aikhenvald, 2013, p. 21).

According to Lichtenberk (2009), the classifier system in Oceanic languages evolved primarily in the domain of alienable possession. "Obviously, a system of possessive classifiers does not have to develop (after all, they are not common in the languages of the world), but if they do develop they are more likely to develop, at least initially, for alienable possession" (Lichtenberk, 2009, pp. 284–85). Supported by experimental research in English (Lichtenberk et al., 2011), he argues that this development was motivated by the fact that possessee nouns that are not inherently relational tend to allow more than one possessive association with possessor nouns. In order to prevent ambiguity, classifiers were gradually invented as a means to disambiguate between different possible readings. This system allows language users to classify possessive relations in terms of how the possessee can be "manipulated" or "owned" so that different classifiers can be recruited with the same possessee (Aikhenvald, 2013, p. 19). For instance, in Mussau-Emira, speakers distinguish linguistically between 'my chicken (domestic animal)' and 'my chicken (to eat),' or likewise, 'my coconut tree' and 'my coconut (to drink)' (Brownie & Brownie, 2007). In this way, the classifier system of Oceanic languages allows speakers to "highlight different uses for the same object" (Aikhenvald, 2013, p. 21), a unique way of expressing alienable possession not found in other languages around the world. Such cross-linguistic relativity, along with the partial overlapping of features, favors the tool hypothesis with regard to language.

Referential Density

Pro-drop ("pronoun dropping") languages are characterized by the omission of free personal pronoun forms in sentences where nevertheless such person- or object reference is implied. As a general cross-linguistic tendency, it has been observed that pro-drop typically occurs in languages with rich inflectional morphology on the verb, which then also includes person/number inflection for subject and/or object referents (Taraldsen, 1978; Huang, 1984). Typical examples include Indo-European languages like Spanish, Portuguese, Italian, and Polish, but pro-drop is also prominent in most other language families around the world. Moreover, as in the case of Japanese, Korean, or Chinese, it occurs in languages that do not show any such inflectional morphology at all (Huang, 1984). On the other hand, typical non-pro-drop languages are those that show intermediate morphological complexity on the verb, that is, languages that partially lack transparent person/number inflection on the verb. In particular, non-pro-drop has been identified as an areal typological feature of the standard varieties of Northern European languages such as French, English, and German (Haspelmath, 2001).

The pro-drop feature can be further extended to referential expressions (NPs) in general, that is, the tendency of some languages to refrain from overt expression of referents by means of either pronouns or lexical noun phrases (e.g., English *he*, *John*, *the boy*). In this context, some interesting microscale typological research has been conducted on what Bickel (2003) refers to as *referential density*. This term is defined as the ratio between possible referential expressions (argument NPs, i.e., nouns or pronouns) within a sentence and the actual realization of such expressions in natural discourse (Bickel, 2003, p. 708). In an experimental study, Bickel (2003) was able to show that even geographically closely affiliated languages like Maithili, Nepali, and

Belhare, all spoken in the Nepalese Himalayas, display quite different degrees of referential density in spoken discourse—despite all three of them being broadly classified as pro-drop. To test this, speakers of all three languages were exposed to the same set of standardized stimuli (Pear Story movie clip; Chafe, 1980) and asked to subsequently narrate their experiences to naïve but familiar subjects in another room. The findings were that the narrated speech of Maithili and Nepali speakers showed a significantly higher referential density than that of Belhare speakers, even though all speakers were exposed to the same displaced conditions. Bickel (2003) observes that in Belhare, "arguments are not even mentioned when they are new: the speaker begins her narrative with saying that someone picked mangos (the locally perceived version of the pears in the original story), but the person doing the picking is neither introduced nor mentioned throughout the paragraph" (Bickel, 2003, p. 709). Moreover, even when new persons appear on the scene in the original story, "this does not prompt the speaker to use an overt NP for this [new] referent" (Bickel, 2003, p. 709). By comparison, Maithili speakers are much less prone to drop NPs, even though in principle "there is no clause structure," even in this language, "where NPs would ever be obligatory syntactically" (Bickel, 2003, p. 709). In particular, Maithili speakers showed greater sensitivity to switch-reference contexts in which new referents were introduced. They would use overt NPs to identify the referents, first by lexical nouns and then by pronouns (Bickel, 2003, p. 710). By contrast: "In the Belhare text [spoken narrative], the speaker is very implicit about referents. Identifying who did what in the story is mostly the listener's task. By relying on verb semantics and general world knowledge, the listener can infer much about referential continuity and discontinuity" (Bickel, 2003, p. 710).

This picture of cross-linguistic variation and family resemblance in referential marking can be further broadened. Merlan and Rumsey (1991) report that the Papuan language Ku Waru (see also the discussion on reciprocity in chapter 5) customarily relies on pro-drop and that, moreover, optional case marking on NPs systematically blurs the distinction between transitive ("directed") and intransitive ("non-directed") event descriptions:

> In longer texts [spoken narratives], any of these NP argument types may be deleted from the clause when its reference is clear from the linguistic context [. . .]. The ergative case postposition is also optional [. . .]. In textual examples where it is absent from the clauses of this type, informants readily restore it in what they treat as equivalent repetitions of the same sentence. (Merlan & Rumsey, 1991, p. 337)

Hence, in addition to frequent argument ellipsis (pro-drop), there is no obligatory formal distinction between transitive and intransitive sentences in this language. The explicit marking of transitive subject NPs as ergative is merely optional, while in many other languages it is obligatory. In comparison with other languages, this yields the effect that utterances do not clearly distinguish between subject (Agent) and object (Patient) on any formal grounds (though verbs still carry person/number agreement and thereby provide some clues). Rather, the distinction rests on semantic-pragmatic grounds and on contextual inferences. According to Rumsey (1990), similar observations pertain to the Australian Aboriginal language Ungarinyin, spoken in the Kimberley region of Western Australia:

> To Europeans, Ungarinyin discourse seems highly elliptical in that, for example, one or more of the entities that are cross-referenced on the verb are usually not referred to by any other element within the clause. But this is not "ellipsis" in Halliday and Hasan's functional sense, since the pronominal elements on the verb do "refer" to these participants, without recourse to any other "wording" by which those participants may have been referred to in any other clauses. (Rumsey, 1990, p. 350)

In this context, Rumsey (p.c., July 2018) adds: "But in Ungarinyin the system of pronominal affixation is very rich, with 4 genders, 4 number and pronominal prefixes showing them for both subject and object. Those play a big part in indication which is which." In other words, speakers of different languages rely on different linguistic tools in order to establish referential identity. Such different means are neither structurally nor functionally fully identical across languages. For instance, the standard European languages seem somewhat deficient in their ability to provide their speakers with sufficient morphosyntactic person/number/gender agreement on verbs in order to establish discourse referents without additional NPs. Many other languages around the world, by contrast, have evolved such morphosyntactic complexities and thereby provide their speakers with economical ways of establishing referents, relying on complex verb formations rather than complex NP structure. These observations are again compatible with the tool hypothesis with regard to language.

The storytelling in Pirahã, a minority language spoken in the Amazonian rainforest, provides yet another example of a how referential density may vary in natural linguistic discourse. From the European perspective, the linguistic references to persons or things are often extremely vague. Informants will begin their narratives with indexical references before having identified the persons referred to. Further specifications only incidentally follow, often after various repetitions of the same initial reference. For example, in a Stolen Brazil Nut Groves story discussed in Everett (2016, pp. 182–84), the informant begins his narrative with "He did not pay Xoágaii," followed by the utterance "He did not pay the Pirahã," and then yet another sentence that still does not provide any clue as to the identity of "he." Only several utterances later does the informant refer to some "foreigner" who is subsequently further identified by name as "Chico," suggesting, only through inferences on the listener's part, that this might be the initial "he." Similarly, Everett (2016) observes, proper names will be introduced without further context: "It is fascinating that he [the informant] introduces (as all Pirahãs do in all texts) proper names of people with no further description, even those (at that time, all the names) I have never met" (Everett, 2016, p. 189). Speakers simply assume, being themselves part of a culture where everyone knows everyone else (in Everett's lingo, a "society of intimates"), that others know whom they refer to (Everett, 2016, p. 191). On the other hand, going beyond referential density in the way defined above, Pirahã stories are riddled with repetitions, a fact Everett ascribes to stylistic emphasis on important contents of a story as well as the noisy environment that makes it difficult to get messages across the first time (Everett, 2016, p. 193). A most impressive example is the Killing the Panther story given in Everett (2008, pp. 123–26), where in one place (as in several others) no less than six utterances in a row contain the basic message that "the panther pounced the dog" and thereby killed it, before the storyteller himself killed the panther by shooting it in a dramatic series of events. This, too, may be considered a special form of referential density on a different scale of natural discourse, one which seems especially characteristic of nonliterate societies.

To conclude, the repertoires of establishing reference to persons and things by linguistic means and the way they are intentionally employed by speakers in ongoing conversation and narrative is a highly culture-specific enterprise. European expectations often fail in respect to cultures outside the grasp of Western civilization. People living in different cultures develop different "fashions of speaking" (Whorf, 1956) based on different ways of interacting that shape their communities at large, and the resulting linguistic repertoires reflect the different backgrounds and constraints that these ideologies create (Rumsey, 1990; Everett, 2016). It is important to realize that these cross-linguistic differences are real and not just variations of some universal cognitive scheme: "Do some societies communicate less or use linguistic communication less than others? It seems so" (Everett, 2017, p. 10). Such cultural relativity defines all genuine tool cultures.

Other Linguistic Phenomena

We need not go into further detail in order to make the claim that many other core-grammatical features linguists describe and analyze submit to the observations made above. Many language-typological investigations have highlighted the fact over the past decades that natural languages differ immensely with regard to the formal and semantic/pragmatic strategies they employ in order to express specific functional categories (Haspelmath, 2007a; Evans & Levinson, 2009). Aside from the examples given above (reciprocals, possessives, referential expressions), other good examples include the marking of tense (Bybee et al., 1994), case marking and grammatical relations (Corbett & Noonan, 2008), differential object marking (Comrie, 1989; Aissen, 2003; Haspelmath, 2008b, pp. 13–14), and the formation of complex sentences (e.g., Hale, 1976; Mithun, 1984; Comrie, 1998; Everett, 2004). In particular, the functionalist analysis highlights the possibility to consider semantic/pragmatic and thus usage-oriented factors as a means to explain why certain cross-linguistic patterns of marking occur and thus how linguistic form relates to meaning and communicative intentionality. The consideration of the tool hypothesis lends further support to such functionalist analyses and adds to our understanding of how form–function relationships in natural languages should be conceptualized, namely in terms of flexible *means–ends dissociations* and characteristic *patterns of variation* that are defining of all genuine tool uses and their cultural-historical origins (see chapter 3). Some works in the functionalist tradition (especially, see Everett, 2012, 2016, 2017) already conceive of natural languages as *cultural tools*. The present study adds to this analysis a direct evolutionary link to the material cultures of nonhuman primates.

THE LARGER CONTEXT OF THE DEBATE

"Since there is a hierarchy in [. . .] an object itself, genuine truth about this object can only be given in a *set* of statements which, by its order, reproduces the hierarchy of the objective structure" (Köhler, 1966 [1938], p. 27; emphasis not added). In the same vein, Köhler continues to observe that it is possible to make "correct" statements about an object of study, while at the same time omitting to specify its proper place within the hierarchy of other "correct" statements about the same object. In this way, single "correct" statements about only minor or even *unessential* features of an object can become ambiguous, "because their meaning lacks the context in which they could become definite and fair in respect to the objective structure" (Köhler, 1966, p. 27). In the worst case, "[i]t follows that single, or a few, 'correct' statements which refer to unessential phases of a subject-matter are, when given alone, not as much ambiguous as positively misleading and, always in the same sense, untrue" (Köhler, 1966, p. 28).

If we wish to make holistic statements about the nature of human languages and human linguistic capacity, we therefore have to choose our set of statements carefully—unless we are satisfied with merely being "correct," rather than "true," with respect to our object of study. The more our single correct statements refer to basic or *essential* features of the object, the more they are allowed to stand on their own, without additional context, and without belying the true nature of "the nucleus of the subject-matter" at hand (Köhler, 1966, p. 27).

Having put forward the claim in this book that languages are proper *tools* of communication, we do aspire to having captured a truly basic and essential feature of human linguistic capacity and the properties of natural languages. In this sense, this work aspires no less than others have done before. Considering the ongoing debate and the plurality of voices participating in it, we want to specify more clearly where within this debate the tool hypothesis stands and how it relates both in positive and negative terms to other mainstream theories within the field.

As we have seen above, the tool hypothesis follows suit with other functional-cognitive approaches that reject the symbolic-computational conception of human syntax and linguistic communication in general (Tomasello, 1995, 1998; Savage-Rumbaugh et al., 1998; Haspelmath, 2008a; Evans & Levinson, 2009; Everett, 2012, 2016, 2017; Ibbotson & Tomasello, 2016). Like functionalism, it does not accept the postulate of the autonomy of syntax and thus the idea that abstract mathematical algorithms operate on abstract symbols as a means to generate meaning and propositional content in the human brain/mind. Together with cognitive and functional linguists, we maintain "that language is not an autonomous 'mental organ,' but rather that it is a complex mosaic of cognitive and social communicative activities closely integrated with the rest of human psychology" (Tomasello, 1998, p. ix). "Symbolic computation" is thus not considered an essential, or even minor, feature of natural languages. In more positive terms, it amounts to the statement that human linguistic capacity and the phenomenological linguistic patterns we observe are "usage-based" and crucially depend on "interaction" between human subjects in complex social-ecological environments in order to emerge and evolve historically (Keller, 1994; Tomasello, 1999; Croft, 2000; Kemmer & Barlow, 2000; Enfield & Levinson, 2006; Tomasello, 2008; Evans & Levinson, 2009; Everett, 2012, 2016; von Mengden & Coussé, 2014; Dor et al., 2014; Ibbotson & Tomasello, 2016; Kuhle, 2018a, 2018b).

Yet, as the past twenty years or so have shown, this term (or statement) *usage-based* still leaves room for considerable ambiguity. Functionalism is not as straightforward as it may seem at first sight (Kolodny & Edelman, 2018). Even the earliest, partly collectively orchestrated initiatives among functional-cognitive linguists and typologists in the 1990s still considered the "symbolic aspect" an essential feature of all human languages. In this view, "all language structures are *symbolic* instruments that serve to convey meaning, from the smallest morphemes to the most complex constructions" (Tomasello, 1998, p. xi; emphasis added). Homo sapiens is "the *symbolic* species," (Deacon, 1997; emphasis added) or "the only species with a *symbolic* language" (Gärdenfors, 2004, p. 237; emphasis added), and "the distinguishing feature of natural languages is its central *symbolic* component" (Lucy, 1996, p. 40; emphasis not added). Such functionalist notions of the "symbolic nature" of human languages are commonly based on the Peircean classification of communicative signals/signs into the three different types of "icon," "index," and "symbol" (Lucy, 1996; Deacon, 1997; Everett, 2016, p. 177; Everett, 2017, pp. 6–7). Symbols are distinguished from the other two types on the grounds that they are based on *social convention* and have *no physical correlation* with the things they refer to in ordinary conversation. In Lucy's own words:

> Following Peirce (1932; see also Benveniste [1939] 1971 and Piaget 1973), a *symbolic* sign involves a sign vehicle which stands for some object *only* by virtue of being so taken to stand for it by an interpretant. In other words, the sign–object relationship is established conventionally in a social group rather than motivated by a "natural" connection as in the case of *icons*, where relations of resemblance obtain between the sign vehicle and its object, or *indices*, where relations of co-presence obtain between the sign vehicle and its object. The signal systems characteristic of other species always depend on iconic or indexical linkages of sign vehicle and object for their effectiveness even when also conventionalized to some degree. Human groups also continue to signal by means of icons and indices, so the presence of symbolic signs does not eliminate the presence or significance of the other sign types. (Lucy, 1996, p. 40; emphasis not added)

At least within mainstream functionalist linguistics, closely associated with language typology, historical linguistics and cognitive linguistics, this continued prominence of the postulated "symbolic" character of language use has had and continues to have a strong impact on how language is defined more narrowly. The most influential consequence of this postulate has been the *social intelligence hypothesis* (see above and the discussion in chapter 2), and hence the claim that "human sociality" and in particular "shared intentionality" play a key role in explaining the emergence of linguistic *symbols* through special modes of human interaction and intersubjectivity (Tomasello, 1999, 2008; Enfield & Levinson, 2006; Zlatev et al., 2008b; Evans & Levinson, 2009; Dor et al., 2014). In addition to defining human linguistic behavior as essentially "usage-based" and "symbolic," it is here also defined as essentially "prosocial" or "ultrasocial" and based on special forms of "we"-modes of interaction and collaboration. Such intelligence is seen as "discontinuous" in relation to the intelligence and behavior of all other animal species, including the great apes.

The present argument, which postulates the homology to prelinguistic tool intelligence, parts company with this core definition of human language. As has been argued elsewhere (Kuhle, 2018a), shared intentionality does not necessarily presuppose intelligent abilities that are entirely absent in prelinguistic animals. It merely requires an *extension* of the kind of *general intelligence* already in place in our closest living relatives and here naturally displayed, among other things, in the physical tool-using manipulations of the environment. In this view, then, languages and human linguistic behavior are not defined as essentially "social" (nor, for that matter, as "symbolic"; more on this point below), but as functionally "tool-like" in character, operating mostly under displaced conditions and always harnessed to the intentions of the individuals and their respective experiences and anticipations with regard to each other and to the linguistic materials employed in any given context of interaction. Functional linguists frequently betray the same kind of assumptions when they draw attention to the "strategic interaction" in linguistic activity (Traugott, 1990, p. 514) or the differential "experiences and assessments" in linguistic "face-to-face interaction" (Hopper, 1987, p. 142). It also becomes foregrounded in typological investigations that conceive of functional explanations in terms of processing difficulty, frequency effects, or various forms of iconicity (Haiman, 1983, 2000; Givón, 1985, 1991; Croft, 2000; Hawkins, 2004; Haspelmath, 2008a, 2008b; Good, 2008).

Intriguingly, this idea of defining human linguistic behavior in terms of *general intelligence*, evolutionarily rooted in prelinguistic intelligence, has been developed before. It formed part of a highly controversial debate on the linguistic capacities of nonhuman primates. As was mentioned in chapter 2, this *ape language debate*, or rather the empirical research that inspired it, was pioneered by the anthropologist Sue Savage-Rumbaugh and her colleagues since the 1970s and gained considerable momentum until well into the late 1990s. In a lucid retrospect account of this debate, Savage-Rumbaugh, Shanker, and Taylor (1998) remark:

> [T]he fact that apes like Kanzi, Panbanisha, Mulika, and Panzee have learned to comprehend and produce simple language when raised in an environment where they are spoken to and treated as competent communicative partners has not had the impact on the fields of linguistics, philosophy, psychology, and behavior that such findings would have had if the results had been accepted as legitimate accounts of ape behavior (Savage-Rumbaugh, Murphy, Sevcik, Brakke, Williams, & Rumbaugh, 1993). (Savage-Rumbaugh et al., 1998, p. 191)

Intriguingly, the authors also hint at the possibility that other empirical approaches to primate intelligence might some day lead to similar findings and be successful in shedding new light:

> Only by getting some insight into the mental world of animals can we break down these conceptual boundaries that we ourselves have erected. While we may eventually discover many ways of doing this, right now we have only one—and that is language. We need to talk to the apes, and we need to listen when they talk back. [. . .] At least we are going to have to do this until we become more proficient at recognizing intelligence by other means. (Savage-Rumbaugh et al., 1998, p. 190)

The current work has sought to do just this. The existence of *material cultures* among nonhuman primates and their predictable patterns of variation (and family resemblances) across social groups opens up a new window into thought and, by drawing the analogy, into language. In many respects, the analysis offered here stands shoulder to shoulder with the notion of general intelligence developed in these earlier works brought forth in the ape language debate. Particularly, the argument sides with the fundamental assumption that human intelligence, just like prelinguistic intelligence in other animal species, is not strictly *hierarchical* and *symbolic* in nature, but rather functionally and structurally flexible and adaptive:

> The kind of intelligence that accepts whatever nature offers from one moment to the next, and tries only to predict what is the most efficient thing to do given the current situation, is what the brain of *Homo* evolved to do for the entirety of our existence on the planet prior to the domestication of plants and animals. [. . .]
>
> As enamored as we are of the hierarchical intelligence that has generated our modern technological advances, it is useful to recognize that this sort of intelligence is a relative latecomer on the scene, and that it may have a limited sort of usefulness. (Savage-Rumbaugh et al., 1998, p. 188)

In this view, hierarchical intelligence as well as modern technological advances characteristic of Western-style civilization are only very special cases of human cultural development that over time have had a rather strong impact on how we now conceive of human nature as a whole.

The reference to technological achievements also reminds us of another prominent position in the current debate. Another way of conceiving language has been to define it as a "social tool" or "social technology" (Dor, 2011, 2014; Everett, 2012, 2016, 2017). As is outlined in Dor (2014), the attribute "social" is taken from the functionalist literature at large and acknowledges the fact that linguistic behavior belongs into "the social realm—between speakers, not in them" (Dor, 2014, p. 105). Moreover, again concurring with other functionalist views, the idea posits a "symbolic landscape" for linguistic reference that consists of "*collective* meaning" negotiated and accumulated through the linguistic community over historical time spans (Dor, 2014, p. 111, p. 113; emphasis not added). But most important, the notion identifies all languages as "socially constructed technologies," or more specifically, "language as a socially constructed technology, dedicated to the bridging of the experiential gaps between its users through the instruction of imagination" (Dor, 2014, p. 124). As is emphasized, "[t]he approach takes the characterization of language as a technology quite literally, and suggests that it must have evolved like other technologies do (Arthur, 2007)" (Dor, 2014, p. 107). This positions language close together with other communication technologies such as the writing system, the telephone, or the Internet (p. 105).

On the face of it, the term *technology* may seem to recommend itself to the tool hypothesis and thus the idea that natural languages consist of complex tool repertoires in the shape of grammatical and other devices. However, the term is again fraught with ambiguity. Ingold (1993) cautions us that there exists a fundamental difference between "tools" and "technologies." The former are flexible extensions of the human body under the continuous control of a subject's emotions and intentions; the latter, by contrast, is defined as autonomous machinery explicitly developed to function *independently* of the human body and *outside* of the individual's control. Technological inventions are thus not at all the same as the manipulative skills learnt and practiced by individuals as a way to interact flexibly and adaptively with the environment.[3] *Technology* is therefore a highly misleading term and is in principle not required as an essential feature of language as an intelligent mode of intentional communication. Rather, it draws attention away from the "adaptive, context-dependent intelligence" highlighted earlier (Savage-Rumbaugh et al., 1998, p. 188). One may try to suppress it, but the first association with technology goes hand in hand with the technological achievements of industrialized societies and their reliance on complex, automatic machinery outside of the immediate control of the individual.[4] It is of course possible to support such an analogy to language, but this is most likely to reinforce the hierarchical and symbolic-computational definition of human linguistic behavior as outlined at the outset of this section. The only difference is that in the "social technology" paradigm, the linguistic conventions are considered "laws in social terms," not in biological-genetic terms (Dor, 2014, p. 123). Either way, the present tool hypothesis does not sustain the notion of language as such an autonomous symbolic system.

The current tool hypothesis, instead, argues for a radical redefinition of language even from a functionalist perspective. It argues that human languages are essentially "non-symbolic." Looking back at the discussion, I myself have been struck at how little the notion of "symbolic reference" has played a role in developing the current argument. Displacement of action and the flexible, constantly negotiable abilities of language users in their reliance on linguistic forms, have figured more prominently in this discussion than anything else. These considerations have also been the source to explaining how "arbitrariness" comes about, namely through context-dependent innovations and subsequent generalization (*means–ends dissociation*). Moreover, the arbitrariness of linguistic means is further amplified through *collective usage*, that is, social transmission within the group, and long-term *historical development*, that is, through many generations of language users who continuously reuse the linguistic material and respond to processing efficiency constraints (e.g., frequency effects, language contact situations) under variable conditions of living within the community.

I conclude, therefore, that the essential features of human linguistic behavior are all derivable from the prelinguistic notion of tool use and there is no need to invoke "symbolic" reference or representation to characterize linguistic structures of natural languages. In this regard, the current work sympathizes with other radical approaches that have been developed over the past decades. *Distributed views of language* have been particularly emphatic about the fact that human languages are not essentially based on a symbolic code (Cowley, 2011b, 2016; Thibault, 2011). As is pointed out succinctly in Cowley and Vallée-Tourangeau (2017), "once brains cease to be imagined as symbol processors, symbols disappear as does the search for co-variance between 'faculties' and models of memory, grammar or vision" (p. 4). Instead, linguists and cognitive scientists subscribing to this view emphasize the importance of "interactivity" and organism-environment relations to arrive at a more holistic definition of human linguistic behavior (Steffensen, 2011, 2017; Cowley & Vallée-Tourangeau, 2017). Under this perspective, language is primarily *coordinated activity* between human bodies (including the mind/brain) and their extended surroundings, which include other human bodies but also other external resources that sustain human action. *Interactivity* is considered more than just "interaction" between self-organized human bodies (Steffensen, 2017, p. 88). Human beings, as indeed other organisms as well, are considered unbounded entities who depend on continuous pico-scale coordinations with their environment in order to be able to think at all. Thinking is action and intelligent action, including intelligent communication, emerges in close coordination with other human bodies and an extended material environment (Steffensen, 2011, p. 192; Cowley & Vallée-Tourangeau, 2017). The fact that humans are able to engage in extended "sense-making," integrating ongoing perception with the "not-here" and the "not-now," in this view explains why human linguistic behavior is *displaced*, and never just completely situated in the moment of speech (Steffensen, 2017, p. 87). In this sense, interactivity identifies language as a *non-local* phenomenon (Steffensen & Cowley, 2010). *Languaging* is another term frequently used to broaden the scope of linguistic behavior beyond the mere uttering of verbal patterns. In Steffensen's own words:

> Human beings engage in "whole-bodied communication" which, in many circumstances, has more to do with the bodily (or body-worldly) co-actions by

> interlocutors than with what they (think they) say. On this view language is a heterogenous set of physical, cognitive and social activities that unfold in real-time on many time-scales. It arises as we adjust to each other and co-ordinate our life-worlds, behaviorally and cognitively. (Steffensen, 2011, pp. 192–93)

Interestingly, this radicalized view of linguistic behavior, which deliberately by-passes the symbolic code view of language and brings aspects into focus that are considered "nonlinguistic" in other theoretical accounts, also fits well with at least some voices within the functionalist debate. Everett (2016), for example—who also promotes the tool notion with regard to language, although not with reference to prelinguistic tool cultures in great apes (Everett, 2016, pp. 104–5, p. 122)—offers a book-length account of "dark matter" as the either unspoken or ineffable subconscious fabric of much of human cultural behavior, including linguistic communication. Like proponents of the distributed view of language, Everett connects the ontology of dark matter with "languaging" and "culturing," and draws on *emicization* as the process of learning by which human beings learn to take "an insider's view" on complex actions that are specific to a particular culture (Everett, 2016, p. 8, p. 26, pp. 29–30). The distributed approach also resonates in other works by linguistic anthropologists who focus on linguistic relativity. In this context, Rumsey (1990, 2010), for instance, draws attention to different "fashions of speaking" (Whorf, 1956) across languages and the connection to different "linguistic ideologies" within different societies. In Standard Average European (SAE), Rumsey argues, language users follow a linguistic ideology that tends to draw "sharp distinctions between words and things, talk and action" (Rumsey, 1990, p. 346). The ideology tends toward the culturally acquired perception that the relationship between words and actions is not directly physical and social. Different wordings can be used to say essentially the same thing. Also, what is not explicitly stated in an utterance cannot be referred to, that is, properly communicated. In other words, wordings represent the world *symbolically* and communication therefore, too, is placed in a kind of virtual reality. Australian Aboriginal ideology, by contrast, rejects this distinction between words and actions. Speaking itself is action and the particular wordings themselves incorporate what is being said. In other words, borne out of a special history, people in Western societies find themselves under special constraints to relate linguistic "wordings" with their environment. This finding by and large agrees with what has been argued elsewhere in the literature on distributed cognition and the observation that a *written language bias* in Western societies has had a strong impact on how the language sciences at large have come to conceive of language as a thoroughly *disembodied* object of study (Linell, 2005; Steffensen, 2011; Cowley, 2016).

To conclude, then, the tool hypothesis offers a useful way of defining language in positive terms that do not draw on "symbolic reference" as an essential feature of natural languages (though certain cultural ideologies may still lead to the use of linguistic means as true symbols, not unlike mathematical operations). It allows us to conceive of linguistic patterns or wordings as culturally evolving tools that emerge in coordinated speaker-hearer interactions that rely on an extended material environment, and an *extended intentionality* that allows humans to integrate information from both the physical and the social world to engage in active social collaboration (Kuhle, 2018a, 2018b). Natural languages are thus no more and no less than the "material cultures" that evolve historically in the domain of intentional communication.

IN A NUTSHELL

"I thus choose to blame all of this complexity on reality—though it is of course possible that we just do not understand everything well enough to find the hidden simplicity" (Tomasello, 2008, p. 319). This work has aimed to uncover at least some aspect of "hidden simplicity." The tool hypothesis discerns a psychological infrastructure that is shared between humans and nonhuman primates—a general intelligence we have characterized throughout this book as *tool intelligence*. It yields testable criteria of cultural development independently of language and allows their testing on natural language data. In a case study on reciprocal constructions in languages around the world, we found that we can verify the hypothesis' predictions for *cultural variation and family resemblance* in cross-linguistic comparison. It thereby provides positive evidence for an explanation of language evolution that rests on assumptions of cognitive continuity in relation to our closest living relatives, the great apes.

NOTES

1. For similar positions see Pinker & Bloom (1990), Chomsky (2002b, pp. 147–48), and Jackendoff (2007).

2. The literature is full of acknowledgments of these facts about cognitive constraints in working memory and thus linguistic performances. However, the underlying theoretical assumptions and thus implications for the ontology of linguistic structures are very different. For example, Hurford (2012) agrees that the production of syntactic forms is constrained by working memory capacities and discourse intentions. However, he maintains that the underlying linguistic form is not explained by these various abilities and intentions of usage. He still keeps to the generative

model and distinguishes between a conceptualizer system and separate grammatical encoder (pp. 522–23). Proponents of the functional (cultural) view of language differ in this regard. For example, Tomasello (2008) consistently assumes direct effects of memory constraints and pragmatic inferencing on the construction of linguistic utterances, and hence argues coherently within a usage-based framework. Yet, he doesn't link such pragmatics of usage with the tool hypothesis. His main focus lies on social intelligence and hence prosocial skills and motivations specific to the human species alone. He thereby postulates another extreme, a kind of social module, as a distinct psychological substrate of linguistic behavior that has no evolutionary foundations in prelinguistic behaviors. The tool hypothesis opposes such discontinuity claims born out of either generative or functionalist theorizing.

3. Some scholars, such as Dan Everett, for instance, will remark that technologies (smartphones, laptops, automobiles, etc.) are oftentimes also just handled like "tools" by subjects (p.c., April 2018). Most of the time, individuals lack complete intelligent insight into the actual design of complex machines but learn pragmatically how to use them for their individual purposes in their daily activities. Moreover, following Everett (2016), *emicization*, that is, getting an insider's perspective and using one's body experience to achieve this, can be a powerful concept of learning in a cultural environment that helps explain how such pragmatic and culture-based acquisition of skilled behavior can take place. However, as outlined in the ensuing main text, I contend that we do not require the notion of "technology" in order to make this point. We may just relate to tools and not to technologies in order to construct a meaningful analogy to linguistic behavior.

4. I thank an anonymous referee for pointing out to me the prominence of the tool concept in Karl Marx's work and his legacy to social and cultural anthropology focusing on "material culture" (Hicks, 2010). Very likely, Ingold (1993) also has some of Marx's ideas in mind when he discusses the distinction between practical, skilled tool uses and automated technology that reduces the human individual to a mere "hand" who presses buttons on a machine. There exists a rich literature on Marx's ideas concerning technological development and its consequences for capitalist societies, which also includes some notion of tools (e.g., MacKenzie (1984), Marx and the Machine, *Technology and Culture, 25 (3)*, pp. 473–502. Adler (1990), Marx, Machines, and Skill, *Technology and Culture, 31 (4)*, pp. 780–812). However, given that Marx's work and the reception thereof is far beyond my expertise and falls outside of the scope of the present argument, I expressly refrain from integrating Marx's ideas into the present work.

Appendix A

Language Information

Table 10.1 List of Languages Systematically Investigated in This Study.

	Language	*Affiliation (Country)*	*References*	*Source Material*
1	Yidiny	Cairns subgroup, Pama-Nyungan (Australia)	Dixon, 1977, 1991	grammar, narratives
2	Dyirbal	Herbert River group, Pama-Nyungan (Australia)	Dixon, 1972	grammar, narratives
3	Dyabugay	Cairns subgroup, Pama-Nyungan (Australia)	Patz, 1991	grammar
4	Ngiyambaa	Central NSW group, Pama-Nyungan (Australia)	Donaldson, 1980	grammar, narratives
5	Jiwarli	Mantharta, Pama-Nyungan (Australia)	Austin, 1997, 1992; Burgman, 2006	grammar sketch, narratives, dictionary
6	Yindjibarndi	Ngayarda, Pama-Nyungan (Australia)	Wordick, 1982	grammar, narratives
7	Martuthunira	Ngayarda, Pama-Nyungan (Australia)	Dench, 1995	grammar, narratives
8	Panyjima	Ngayarda, Pama-Nyungan (Australia)	Dench, 1991	grammar, narratives
9	Nunggubuyu	Arnhem Land group, Non-Pama-Nyungan (Australia)	Heath, 1980, 1981, 1984	grammar, narratives, dictionary
10	Nyulnyulan languages	Non-Pama-Nyungan (Australia)	McGregor, 1999	paper on reciprocals
11	Kâte	Eastern Huon, Finisterre-Huon (Papua New Guinea)	Keyßer, 1969/25; Pilhofer, 1969/33; Flierl & Strauss, 1977	grammar, dictionary

(*Continued*)

Table 10.1 (Continued)

	Language	*Affiliation (Country)*	*References*	*Source Material*
12	Ku Waru	Chimbu-Wahgi, Trans-New Guinea (Papua New Guinea)	Merlan & Rumsey, 1991, 1989	grammar sketch, word list, narratives
13	Skou	Skou, Macro-Skou (New Guinea)	Donohue, 2004	grammar, narratives
14	Kilivila	Oceanic (Trobriand Island, Papua New Guinea)	Senft, 2011	paper on reciprocals, video-clip material
15	Mah Meri	Aslian (Malay Peninsula)	Kruspe, 2011	paper on reciprocals, video-clip material
16	Jahai	Aslian (Malay Peninsula)	Burenhult, 2011	paper on reciprocals, video-clip material

The table includes information about the linguistic affiliation (language family) and geographic location (country) of each single language, as well as the primary references and type of language material (grammar, dictionary, text collection) relied upon for the analysis of reciprocal marking patterns. If not taken from the respective references, the affiliations are taken from Dixon (2002) or Ross (2005), respectively. In the right-hand column, *grammars* refer to comprehensive linguistic descriptions of a language with sections on phonology, morphology, lexicon, and syntax; *narratives* refer to spoken language narratives of language informants, which have been transcribed, glossed, and translated by linguists; *dictionaries* are comprehensive descriptions of lexical inventories of a language.

The language name in no. 10 indicates that the analysis did not rely on a single language but a whole group of languages belonging to the same family (Nyulnyulan languages); in addition, the analysis did not (as in most other cases) rely on primary references, but on a detailed synopsis on reciprocal marking given in McGregor (1999). Another exception poses the language materials of the Austronesian language Kilivila (no. 14) and the Aslian languages Mah Meri (no. 15) and Jahai (no. 16), respectively. The data is here taken from "papers on reciprocals" (presented in edited volumes) that were in turn based on elicitation tasks involving video clips specifically devised for the elicitation of reciprocals (Evans et al., 2004; Majid et al., 2011b).

Data sources are always taken into account when discussing the material in the main text (chapters 5, 6, 7, and 8).

Appendix B

Language Data Sets

NUNGGUBUYU

The six tables below, Tables 11.1.1–11.1.6, list reciprocal predicates with verbal *-nyji* in Nunggubuyu according to event-type category. The entire data set is based on Heath (1981) (only in some rare cases also Heath, 1984), which lists dictionary entries in alphabetical order and cross-references to text examples in Heath (1980). The distinction between "reciprocals proper" and "lexicalized reciprocals" follows the distinction between semantically and formally transparent (= grammatical) as opposed to idiomatic (= lexical) predicate formations (Nedjalkov, 2007b). The tables are complementary to Figure 5.1 in chapter 5.

Table 11.1.1 Reciprocal Use of Verbal Suffix *-nyji* in Nunggubuyu ("antagonistic action") (Heath, 1980, 1981).

	Reciprocals proper:		*Base:*	*Source*
1	*=ngawi-'-jga-nyji-* die-?-Caus-Recip 'to cause each other to die'	<	*=ngawi-'-jga-* vtr. 'to cause to die'	H, 1981: 142
2	*=ri-nyji-* spear-Recip 'to spear each other' (Also: 'to get into a fight')	<	*=ra-* vtr. 'to throw a spear at s.o.'	H, 1981: 150
3	*-nyi=ra-wi-jga-nyji-* curse=sing-?-Caus-Recip 'to curse each other, exchange curse songs'	<	*-nyi=ra-wu-* vtr. 'to curse (s.o.) by singing *nyiri* curse song'	H, 1981: 161
4	*=adhuga-nyji-* stab-Recip 'to be stabbing each other'	<	*=adhuga-* vtr. 'to stab (s.o.)'	H, 1981: 172
5	*=wargalhi-nyji-* do.sth.repeatedly-Recip 'to do sth. repeatedly to each other'	<	*=wargalha-* vtr. 'to do sth. repeatedly to (person); especially to spear and shoot repeatedly'	H, 1981: 215
6	*=wi-nyji-* hit-Recip 'to hit each other' (Also: 'to fight (each other)')	<	*=wu-* vtr. 'to hit, kill, shoot sb.'	H, 1981: 235
7	*=wa-ngi-nyji-* hit-E:Nasal-Recip 'to hit/kill each other' (Also: Refl 'to hit onself')	<	*=wa-* vtr. 'to hit, kill sb.; also: to bite'	H, 1981: 168
Lexicalized reciprocals				
8	*-a-ngu=bura-nyji-* Ben-E:Nasal=sit-Recip 'to mutually propose (fight)' (Also: 'to put sth. down for each other')	<	*=bura-* vtr. 'to put down, to put (in a place)'	H, 1981: 14
9	*-manda=ngunymaa-nyji-* Group=examine-Recip 'examine each other, seize each other up'	<	*=ngunymaa-* vtr. 'to examine (objects)'	H, 1981: 147
10	*=ri-nyji-* spear-Recip 'to get into a fight' (Also: 'to spear each other')	<	*=ra-* vtr. 'to throw a spear at s.o.'	H, 1981: 150
11	*=wi-nyji-* hit-Recip 'to fight (each other)' (Also: 'to hit each other')	<	*=wu-* vtr. 'to hit, kill, shoot sb.'	H, 1981: 235

Table 11.1.1

	Reciprocals proper:		*Base:*	*Source*
12	*=ja:du-ga-nyji-* finish-Fact-Recip 'to finish each other off; to kill each other'	<	*=ja:du-ga-* vtr. 'to finish (it) up'	H, 1981: 42
13	*=jurjurga-nyji-* push.forward-Recip 'to push/shove each other'	<	*=jurjurga-* vtr. 'to push or drive (fish) forward'	H, 1981: 48
14	*-a=na-nyji-* Ben=see-Recip 'to confront each other (in context of fighting)'	<	*-a=na-* vtr. 'to see s.o.' (?) (< *=na-* 'to see (object)')	H, 1981: 120
15	*-walgara=ra-wa-nyji-* Mult=swear-Fact-Recip 'to all wail at each other'	<	*=ra-* vtr. 'to swear at, to quarrel loudly with'	H, 1980: T71.16.1, p. 322
16	*=argi-nyji-* pull-Recip 'to race with/against each other'	<	*=argi-* vtr. 'to pull'	H, 1981: 223

Source: Author created.

Table 11.1.2 Reciprocal Use of Verbal Suffix *-nyji* in Nunggubuyu ("joint action") (Heath, 1980, 1981).

	Reciprocals proper:		*Base:*	*Source*
17	*-a-ngu=bura-nyji-* Ben-E:Nasal=sit-Recip 'to put sth. down for each other') (Also: 'to mutually propose (fight)')	<	*=bura-* vtr. 'to put down, to put (in a place)'	H, 1981: 14
18	*=da:-'-nyji-* copulate.with-?-Recip '(man and woman) to make love'	<	*=da-* vtr. '(man) to copulate with (woman)'	H, 1981: 22
19	*-a=dharga-nyji-* Ben=send-Recip 'to send (sth.) to each other' (Also: 'to pass/distribute to (sth.) e.o.')	<	*-a=dharga-* vtr. 'to send (sth.) to (s.o.)' (< *=lharga-* vtr. 'to send')	H, 1981: 64
20	*=lharma-nyji-* chase-Recip 'to chase each other'	<	*=lharma-* vtr. 'to chase, follow'	H, 1981: 64
21	*-a=lharma-nyji-* Ben-chase-Recip 'to perform (songs/rituals) for each other'	<	*=lharma-* vtr. 'to perform (ritual)'	H, 1981: 64

(*Continued*)

Table 11.1.2 (Continued)

	Reciprocals proper:		*Base:*	*Source*
22	*-a=jiga-lharma-nyji-* Ben-fire=follow-Recip 'to follow (meet up with) each other by heading for each other's fires'	<	*-a-jiga=lharma-* vtr. 'to follow or track down by following the sign of fires'	H, 1981: 65
23	*=maga-nyji-* tell-Recip 'to tell each other, speak to each other'	<	*=maga-* vtr. 'to tell (sth.) to (s.o.)'	H, 1981: 84
24	*=manma-nyji-* help-Recip 'to cooperate with each other'	<	*=manma-* vtr. 'to help (s.o.), to cooperate'	H, 1981: 97
25	*=marbuy-(w)a-nyji-* knowing-Fact-Recip 'to inform/teach each other'	<	*=marbuy-(w)a-* vtr. 'to inform/teach (s.o.)'	H, 1980: T69.13, p. 313
26	*=na-nyji-* see-Recip 'to see each other'	<	*=na-* vtr. 'to see (object)'	H, 1981: 120
27	*-a-jiga=na-nyji-* Ben-fire=see-Recip '(two or more groups) to see each other's fires in the distance'	<	*-a-jiga=na-* vtr. 'to see s.o. indirectly by seeing fire in the distance'	H, 1981: 120
28	*-a-danggal=ngarayi-nyji-* Ben-gaze=conceal-Recip 'to conceal/shield from one another' 'to avoid (eye contact with) each other'	<	*-a-danggal=ngaraya-* 'to conceal from s.o.'; 'to avert one's gaze (eye contact) from s.o.'	H, 1980: T76.2.3, p. 340
29	*=ragu-nyji-* search-Recip 'to search for each other'	<	*=ragu-* vtr. 'to search for (person, object)'	H, 1981: 154
30	*-a=gadi-nyji-* Ben=shout-Recip 'to shout to each other'	<	*-a=gada-* vtr. 'to shout to (person)'	H, 1980: T47.2.5, p. 244
31	*=wali-nyji-* paint-Recip 'to paint each other'	<	*=wala-* vtr. 'to paint'	H, 1981: 190
32	*=wargu-nyji-* carry-Recip 'to carry each other (on the back)'	<	*=wargu-* vtr. 'to carry sth. on one's shoulder'	H, 1981: 223
33	*-ba=burdhi-nyji-* close.eyes-Recip 'to close each other's eyes'	<	*-ba=burdha-* 'to cause (person) to close his/her eyes'	H, 1981: 252

Table 11.1.2

	Reciprocals proper:		*Base:*	*Source*
34	*=yami-jga-nyji-* say.this-Caus-Recip 'to say this to each other' (Also: Refl 'to say this to oneself')	<	*=yami-jga-* vtr. 'to say this to s.o.'	H, 1981: 263
35	*=yambi-nyji-* speak-Recip 'to speak to each other'	<	*=yambi-* vtr. 'to speak to (person)'	H, 1981: 264
36	*-yany=janga-nyji-* hear.voice-Recip. 'to hear each other's voices'	<	*-yany=janga-* 'to hear the voice of s.o.'	H, 1980: T47.3.5, p. 245
37	*=yi-nyji-* 'to give sth. to each other; exchange' (Also: 'to share')	<	*=yi-* vtr. 'to give (sth.) to (s.o.)'	H, 1981: 271
Lexicalized reciprocals:				
38	*-a=dharga-nyji-* Ben=send-Recip 'to pass/distribute to (sth.) each other' (Also: 'to send (sth.) to each other')	<	*-a=dharga-* vtr. 'to send (sth.) to (s.o.)'	H, 1981: 64
39	*=yi-nyji-* give-Recip 'to share' (Also: 'to give sth. to each other; exchange')	<	*=yi-* vtr. 'to give (sth.) to (s.o.)'	H, 1981: 271
40	*=gaḻij-ga-nyji-* fun-Fact-Recip 'to play with each other'	<	*=gaḻij-ga-* vtr. 'to make fun of s.o.'	H, 1981: 32
41	*-anyji=lharma-nyji-* Com=chase-Recip 'to be involved in a chase'	<	*=lharma-* vtr. 'to chase, follow'	H, 1981: 64
42	*-ṟanggal=na-nyji-* see(sleeping person)-Recip 'to look each other straight into the face'	<	*-ṟanggal=na-* vtr. 'to see (sleeping person)'	H, 1981: 120

Source: Author created

Table 11.1.3 Reciprocal Use of Verbal Suffix *-nyji* in Nunggubuyu ("psychological attitude") (Heath, 1980, 1981).

	Reciprocals proper:		*Base:*	*Source*
43	*=nganybandi-nyji-* want-Recip 'to want (to be with) each other'	<	*=nganybanda-* vtr. 'to like, enjoy, want'	H, 1981: 136
44	*-lhawu=nganybandi-nyji-* words=want-Recip 'to want (to hear) each other's words'	<	See above.	H, 1981: 136
45	*-a=diyaldhi-nyji-* Ben=be.angry-Recip 'to get angry at each other'	<	*a=diyaldha-* vtr. 'to become angry at s.o.'	H, 1981: 164
Lexicalized reciprocals				

Source: Author created

Table 11.1.4 Reciprocal Use of Verbal Suffix *-nyji* in Nunggubuyu ("(dis)similarity") (Heath, 1980, 1981).

	Reciprocals proper:		*Base:*	*Source*

Lexicalized reciprocals				
46	*=ngunyju-ga-nyji* same-Fact-Recip 'to do the same thing to one another'	<	*=ngunyju-ga-* vtr. 'to make (sb./sth.) similar or equal'	H, 1981: 147

Source: Author created

Table 11.1.5 Reciprocal Use of Verbal Suffix *-nyji* in Nunggubuyu ("physical action/state") (Heath, 1980, 1981).

	Reciprocals proper:		*Base:*	*Source*
47	*=dhabid̲bi-nyji-* connect.together-Recip 'to join/connect together'	<	*=dhabid̲bu-* vtr. 'to tie up, connect together'	H, 1981: 15
48	*=r̲aganma-nyji-* face-Recip 'to face each other'	<	*=r̲aganma-* vtr. 'to head toward, to face, to be opposite from'	H, 1981: 153
49	*=r̲uduga-nyji-* be.parallel-Recip '(two objects or rows) be parallel'	<	*=r̲uduga-* vtr. '(one long object or row of objects) to be parallel to (another)'	H, 1981: 164
50	*=wagar̲i-nyji-* head.for-Recip 'to head for each other, converge'	<	*=wagar̲a-* vtr. 'to head for (place, person)'	H, 1981: 179
51	*=wa:li-nyji-* bring.together-Recip 'to come together; assemble'	<	*=wa:la-* vtr. 'to bring (people) together'	H, 1981: 184
52	*=yardi-nyji-* touch-Recip 'to touch each other' (Also: 'to be closely bunched together')	<	*=yardi-* vtr. 'to touch'	H, 1981: 269
53	*=yali-nyji-* mix-Recip 'to mix honey and water'	<	*=yala-* vtr. 'to mix (often honey) with water'	H, 1981: 260
54	*=yal̲giwa-nyji-* go.past-Recip 'to pass each other'	<	*=yal̲giwa-* vtr. 'to go past (s.o./sth.)'	H, 1981: 262
Lexicalized reciprocals				
55	*=yardi-nyji-* touch-Recip 'to be closely bunched together' (Also: 'to touch each other')	<	*=yardi-* vtr. 'to touch'	H, 1981: 269
56	*=ma-nyji-* get-Recip 'to gather/come together'	<	*=ma-* vtr. 'to get; to pick up'	H, 1981: 80
57	*-yiga=ma-nyji-* fire=get-Recip '(fires) to join together'	<	See above.	H, 1981: 81
58	*=lhangarma-nyji-* reach-Recip 'to meet/run into each other'	<	*=lhangarma-* vtr. 'to reach, overtake, find'	H, 1981: 60
59	*=munduwa-nyji-* examine.closely-Recip 'to come together; assemble'	<	*=munduwa-* vtr. 'to examine closely (and divide into piles); to count or sort objects'	H, 1981: 114

(*Continued*)

Table 11.1.5 (Continued)

	Reciprocals proper:		*Base:*	*Source*
60	*-ḻan=baṉaga-nyji-* knee=hold-Recip 'to hold each other's knees together'	<	*=waṉaga-* vtr. 'to hold, grasp'	H, 1981: 204
61	*-manda=baṉaga-nyji-* group=hold-Recip '(group) to stay together, to hang together'	<	See above.	H, 1981: 204
62	*=a:ru-nyji-* abandon-Recip 'to abandon each other, disperse'	<	*=a:ru-* vtr. 'to abandon, leave behind'	H, 1981: 224
63	*=wula-gi-nyji-* two-Fact-Recip '(two objects) to be paired, to be together'	<	*=wula-ga-* vtr. 'to do sth. twice (to object)'	H, 1981: 240
64	*-mum=bula-gi-nyji-* feet-two-Fact-Recip 'to put feet together (two or more persons)'	<	See above.	H, 1981: 240
65	*=wuruma-nyji-* buzz.around-Recip 'to all swarm together'	<	*=wuruma-* vtr. '(insect) to buzz around (person)'	H, 1981: 254
66	*janda- =yi-nyji-* back give-Recip 'to turn one's back on each other'	<	*janda* nadj. 'back' + *=yi-* vtr. 'to give'	H, 1981: 271; H, 1980: T75.25, p. 339

Source: Author created

Table 11.1.6 Reciprocal Use of Verbal Suffix *-nyji* in Nunggbuyu ("social kinship") (Heath, 1980, 1981, 1984).

	Reciprocals proper:		*Base:*	*Source*
67	*=alguṟ-ma-nyji-* relative-Inch-Recip '(two or more persons) to be related to each other'	<	*=alguṟ* nadj. 'kinsman, relative'	H, 1981: 188
68	*=guḻgu-wa-nyji-* spouse-Fact-Recip 'to marry each other'	<	*=guḻgu-wa-* vtr. 'to marry'	H, 1981: 39; H, 1984: 239
69	*=muruyung-ga-nyji-* elder.brother-Fact-Recip '(X and Y) to be brothers'	<	*muruyung* nom. '(my/ our) elder brother/sister'	(H, 1981: 119) H, 1984: 239

Table 11.1.6

	Reciprocals proper:		*Base:*	*Source*
Lexicalized reciprocals:				
70	=*lhama-nyji-* beget-Recip 'to beget each other (constitute a patriline)'	<	=*lhama-* vtr. '(man) to beget (child)'	H, 1981: 57
71	=*yaba-nyji-* bear-Recip 'to give birth to each other (form a matriline)'	<	=*yaba-* vtr. '(woman) give birth to (child)'	H, 1981: 257

Source: Author created

KÂTE

The tables following, Tables 11.2.1–11.2.3, are complementary to Figure 5.2 in chapter 5.

Table 11.2.1 Simple Reciprocal Verbs ('Rdp-V + e-') in Kâte (Keyßer, 1969 [1925]; Pilhofer, 1969 [1933]; Flierl & Strauss, 1977).

No.	*Reciprocal predicate*	*Interpretation*	*Base (<)*	*Source*
1	*qa-qac e-* Rdp-hit do-	'to beat each other, to fight with each other, to make war'	*qazo* vtr. 'to beat, hit, fight, kill sb./sth.'	K, 373; F&S, 279
2	*â-âkec e-* Rdp-spear do-	'to spear one another'	*âeckezo* vtr. 'to prick, sting, stab, spear sb.'	P, 99; K, 4; F&S, 9
3	*ma-masiri e-* Rdp-tease do-	'to banter, perform sham-fights; to tease one another, play games'	*masiriezo* vintr. 'to banter' (*mamasiri* n. 'jest, joke, play, games')	K, 264; F&S, 189
4	*ʒâ-ʒâmuc e-* Rdp-destroy do-	'destroy one another (enemies)'	*ʒâmuzo* 'to sharpen sth. (knife), to rub sth.'	K, 520; F&S, 382
5	*na-nanʒoŋ e-* Rdp-destroy do-	'to destroy, ruin one another'	*nanʒoŋ jopazo* vtr. 'to destroy them (enemies)'	K, 295; F&S, 228
6	*he-hesuec e-* Rdp-push do-	'to push one another away'	*hesuezo* vtr. 'to push sb. away, to kick sb. away'	F&S, 130, 136
7	*qâ-qâtipec e-* Rdp-push.away do-	'to push one another away'	*qâtipezo* vtr. 'to fling, kick or hurl sth. away'	F&S, 287, 292

(*Continued*)

Table 11.2.1 (Continued)

No.	*Reciprocal predicate*	*Interpretation*	*Base (<)*	*Source*
8	*ki-kic e-* Rdp-bite do-	1. 'to bite or sting each other' 2. 'to meet' 3. 'to close (the ends of a ring)'	*kizo* 'to bite, sting, press sb. or sth., weigh down'	K, 213; F&S, 175, 182
9	*qa-qalac e-e-* Rdp-scratch Rdp-do-	'to scratch each other'	*qalackezo* 'scratch sth., hold sth. with the claws'	K, 337; F&S, 264
10	*fâ-fârec e-* Rdp-drag do-	'to drag each other about'	*fârezo* 'to drag, to pull, to draw near'	K, 81
11	*ha-hanec e-* Rdp-scold do-	'to quarrel'	*haneckezo* 'to scold sb.'	P, 88; K, 139
12	*la-laec e-* Rdp-abuse do-	'to abuse each other'	*lae nalezo* 'to abuse me'	K, 221
13	*e-ec e-* Rdp-hurt do-	'to hurt each other, do sth. to each other, challenge each other'	*ecnezo* 'to hurt him, to do sth. for or against him,	K, 75; F&S, 70
14	*a-analic e-* Rdp-contest do-	'to fall out about sth., contest each other's right to sth.'	---	K, 1; F&S, 1
15	*qe-qesoc e-* Rdp-abuse do-	'to abuse one another, run each other down'	---	K, 378
16	*hu-huc lâ-lâliec e-* Rdp-jump ? do-	'to try to overcharge, surpass each other'	*huckezo* vtr. 'lit. to bring sth. below, to climb/jump over sth.'	K, 164; F&S, 147
17	*si-li-lic e-* Rdp-overtake do- (rdp. of 2nd syllable)	'to overtake one another (e.g. in a race); to outdo one another'	*silickezo* 'to outrun, to overtake sb.; to surpass sb. (in height)'	K, 418
18	*i-iqic e-* Rdp-visit do-	'to visit each other'	*iqickezo* vtr. 'to visit sb., to tarry' (*iiqic* n. 'the visiting of one another')	F&S, 152, 154
19	*he-hetiric e-* Rdp-fail.to.meet do-	'to fail to meet each other, miss one another, leave each other, become separated (by death)'	*hetirizo* vtr. 'to leave sb. behind, separate, be separated, be left out'	F&S, 130, 138
20	*he-hewic e-* Rdp-meet do-	'to meet with a rebuff from one another'	---	F&S, 130
21	*mâ-mâzoc e-* Rdp-take.leave do-	'to take leave of each other, to say goodbye to each other'	*mâzockezo* 'to take leave of sb., say goodbye'	K, 267

Table 11.2.1

No.	*Reciprocal predicate*	*Interpretation*	*Base (<)*	*Source*
22	*ho-honec e-* Rdp-see do-	'to see each other'	*honezo* 'to see sb. or sth.'	P, 88; K, 160
23	*ho-honeliŋ e-* Rdp-look do-	'to look at each other with pleasure'	*honeliŋkezo* 'to look at sth. with pleasure'	K, 160
24	*ho-honefec e-* Rdp-observe do-	'to observe each other'	---	K, 160-61
25	*qâ-qâuac e-* Rdp-touch do-	'to touch each other'	*qâuazo* vtr. 'to touch, stir up, graze against sb.'	F&S, 292
26	*so-sopec e-* Rdp-insult do-	'to abuse, insult each other'	*sopeckezo* 'to abuse, insult sb.'	P, 99; K, 436
27	*fe-fec e-* Rdp-climb.up do-	1. 'to outdo each other in cursing, abusing each other; to upbraid e.o., to vie with e.o.' 2. 'to fall, to tumble one over the other'	(*fezo* vintr. 'to go up, climb up')	K, 90; F&S, 84-85
28	*ho gu-gumec e-* Rdp-dirt.feed do-	'to feed each other with dirt, i.e. offend each other by saying *honane nâc!* (eat my dirt!) etc.'	*ho gumezo* 'to feed sb. with dirt'	K, 126, 157
29	*he-hejac e-* Rdp-reproach do-	'to reproach one another bitterly'	*hejacnezo* 'to beat sth. flat, i.e. to flatten sth. (e.g. with a hammer)'	K, 181; F&S, 130
30	*mu-ʒâ-ʒâic e-* speak-Rdp-hurt do-	'to insult one another'	*ʒâickezo* 'to ache, to hurt'; *muzo* vtr. 'to say, speak sth.'	K, 290, 516, 284
31	*su-surec e-* Rdp-degrade do-	'to degrade one another'	*surezo* vtr. 'to degrade, despise, abuse sb./sth.' (*susure* n. 'abuse, degradation')	F&S, 343
32	*ma-manac e-* Rdp-understand do-	'to come to an understanding with one another'	*manazo* vtr. 'to agree, understand' (*mamanac* n. 'mutual agreement')	F&S, 188, 193
33	*ŋu-ŋuluŋ e-* Rdp-grunt do-	'to grunt at each other (e.g. pigs, but also said of human beings)	*ŋuluŋkezo* 'to grunt (pigs), to moan (people)'	K, 317

(*Continued*)

Table 11.2.1 (Continued)

No.	*Reciprocal predicate*	*Interpretation*	*Base (<)*	*Source*
34	*qo-qoŋgiŋ e-* Rdp-pass.on do-	'to inform one another, to pass sth. on, to hand sth. around'	*qoŋgiŋkezo* vtr. 'to hand sth. on, transfer (a load)'	K, 385; F&S, 298
35	*qa-qazuc e-* Rdp-learn do-	'to practice sth. together, (to teach each other)'	*qazuzo* 'to smash sth., to learn, to practice sth.	K, 374
36	*ma-masuc e-* Rdp-pout do-	'to be angry with one another'	*masulezo* 'to pout, to grumble, to be sulky'	K, 265, 264
37	*qâ-qârâŋgeŋ e-* Rdp-support do-	'to strengthen one another (in an attitude)'	*qârâŋgeŋkezo* vtr. 'to hold sth. rigid/in proper position; to support'	F&S, 286, 287
38	*ŋi-ŋiec e-* Rdp-help do-	'to help each other fill the nets'; 'to help each other'	*ŋiezo* 'to put sth. into a carrying-net or a net-bag'	K, 313
39	*ʒi-ʒimic e-* Rdp-assist do-	1. 'to assist each other in closing the cracks of the canoe'; 2. 'to help, to assist each other, to stand by each other; to unite (against the enemies)'	*ʒimickezo* 1. 'to close sth. with *ʒimic*-resin (wax), e.g. the cracks of a canoe; 2. to close sth. up; 3. to assist, support sb.'	K, 537-38
40	*ʒu-ʒuhuc e-* Rdp-join do-	'to be united, to be joined together'	*ʒuhuckezo* vtr., vintr. 'to add, attach, join, tie, continue sth., to be added to, joined to'	F & S, 398, 396
41	*ba-bafic e-* Rdp-help do-	'to help each other'	*bafickezo* 'to help, assist sb. (work, fight)' (*babafic* n. 'mutual help')	K, 15; P, 99
42	*he-hesanaŋ e-* Rdp-stick.together do-	'to stick together, stand shoulder to shoulder, to support a common cause'	*hesanaŋkezo* vtr. 'to make sth. firm by beating; stick together'	F&S, 129; K, 400
43	*dedec e-, dedehotac e-* Rdp.stick.together do-	intr. 'to stick together; also: to stick, be sticky'	---	K, 58
44	*he-heʒakiec e-* Rdp.strengthen do-	'to strengthen one another, stand firmly together'	*heʒakiezo* vtr. 'to cling to sb. or sth.; to fasten sth. by knocking'	F&S, 130; K, 518

Table 11.2.1

No.	Reciprocal predicate	Interpretation	Base (<)	Source
45	*ke-keteŋ e-* Rdp-encourage do-	'to encourage one another, strengthen each other'	*keteŋkezo* 'to fix sth. with wedges (e.g. a pole in the ground)' (*keteŋ* n. 'a little wedge, hold, support, stay') (*keketeŋ* adv. 'close together')	F&S, 172-73; K, 205
46	*qâ-qârândiŋ e-* Rdp-roam do-	'to roam about together'	*qârândiŋkezo* vtr. 'to knock sth. over, to roll sth. (eg. a log)'	F&S, 286, 287
47	*ʒâ-ʒâfic e-* Rdp-unite do-	1. 'to unite, to enter into an alliance' 2. 'to become entangled in a quarrel, to become engaged in a fight, to fight stubbornly'	*ʒâfickezo* vtr. 'to tie a knot, to bind together (e.g. two strings)' (*ʒâʒâfic* n. 'knot, tie; union, alliance (of two friends, tribes etc.))	K, 515; F&S, 387
48	*u-umbiŋ e-* Rdp-protect do-	'to protect each other, each other's backs'	*umbiŋkezo* 'to wrap up, to envelop sth., to protect sb.'s back'	K, 491
49	*ŋe-ŋezic e-* Rdp-protect do-	'to defend, protect one another'	*ŋezicke zo* 'to protect, to assist sb.'	P, 88; K, 311
50	*so-soŋaŋ e-* Rdp-watch do-	'to control each other, to watch one another'	*soŋaŋkezo* 'to guard, to watch over sb.'	K, 435; F&S, 340
51	*ge-geoc e-* Rdp-replace do-	'to replace one another, to relieve one another'	*geockezo* vtr. 'to represent sb., be sb.'s replacement'	K, 117; F&S, 104
52	*hâmo-du-duc e-* die-Rdp-bury do-	'to die and bury each other'	*hâmozo* 'to die, to be dead'; *duckezo* vtr. 'to bury sb. or sth.'	K, 66, 137
53	*â-âsic e-* Rdp-carry do-	'to carry one another on the back'	*âsickezo* vtr. 'to carry sb. or sth. on the back' (*ââsic* n. 'the act of carrying sb. or sth. on the back)	K, 11; F&S, 9, 15
54	*ba-bac e-* Rdp-hold do-	1. 'hold each other by the hand' 2. 'to marry each other'	*bazo* 'to touch, to hold, to carry in the hand, to do, to put a hand to sth.'	K, 30

(*Continued*)

Table 11.2.1 (Continued)

No.	*Reciprocal predicate*	*Interpretation*	*Base (<)*	*Source*
55	*hâ-hâpoc e-* Rdp-embrace do-	'to embrace each other'	*hâpozo* 'to embrace sb. or sth.'	K, 142
56	*hâpo-mu-muduc e-* embrace-Rdp-kiss do-	'to embrace and kiss one another'	*hâpozo* 'to embrace sb.' *muduckezo* 'to kiss sb.'	K, 142; F&S, 218
57	*gi-gic e-* Rdp-copulate do- *gi-gic ba-bac e-* Rdp-cop. Rdp-do do-	'to copulate (with each other)'	*gizo* 'to copulate with sb.'	K, 122
58	*bâ-bâfuac e-* Rdp-meet do-	'to meet each other'	*bâfuazo* vtr. 'to meet sb.' (*bâbâfuac* n. 'the mutual encounter')	F&S, 29, 32
59	*(faio) qâ-qâruc e-* (on top) Rdp-rush do-	'to fall over each other (e.g. people in a panic)'	*qârucnezo* vintr. 'to rush away' trans. 'to beat off sb., to put sb. to flight' (*faio* 'on top')	K, 357, 78
60	*he-heuac e-* Rdp-touch do-	'to knock against one another in passing'	*heuazo* vtr. 'to touch sb., to tap another in passing'	F&S, 138
61	*ti-tifaŋ e-* Rdp-throw.out do-	'to distribute sth. (things, money etc)'	*tifaŋkezo* 'to throw up, to throw sth. out (eg. a volcano throws upearth)'	K, 471
62	*si-sifuc e-* Rdp-mixed.up do-	'to get mixed up'	*sifuckezo* 'lit. to go opposite to the direction, to be uncertain about sth., to be lost (a track), to be missing, to comingle'	K, 416
63	*mu-muduc e-* Rdp-lie.on.top do-	'to lie on top of one another, to fit tightly'	*muduckezo* 'to lie on sth. or sb.; also: to kiss sb. (mouth upon mouth)'	F&S, 221, 218; K, 281
64	*i-itic e-* Rdp-fit do-	'to fit together (eg. parts, pieces)'	*iticnezo* 'to join, attach sth.' (*iitic* n. 'the fitting together')	F&S, 152; K, 177
65	*ha-hatackezo* Rdp-trace.line.do-	'to fit two lines or edges together exactly, eg. when folding paper'	*hatackezo* 'to trace a line on sth.'	K, 144
66	*he-hefaric e-* Rdp-quarrel do-	'to contest sth., to quarrel over sth.'	*hefâriezo* vintr. 'to quarrel about sth.'	F&S, 127-28

Source: Author created.

Table 11.2.2 Adjunct Reciprocal Verbs ('Adj + Rdp-V + e-') in Kâte (Keyßer, 1969 [1925]; Pilhofer, 1969 [1933]; Flierl & Strauss, 1977).

No.	*Reciprocal predicate*	*Interpretation*	*Base (<)*	*Source*
1	*[okac ba-bac] e-* Rdp-take.revenge do-	'to take revenge on one another'	*okac bazo* 'take revenge' (*okacne* n. 'retaliation, revenge etc.')	F&S, 253 (P, 99; K, 319)
2	*[upeo ba-bac] e-* Rdp-quarrel do-	'to start a quarrel amongst each other'	*upe* n. 'neck, nape of the neck; fig.: conscience' *bazo* vtr. 'to touch, hold, carry in hand, do etc.'	K, 493, 30; F&S, 362
3	*[upeo lo-loc] e-* Rdp-collar do-	'to collar each other'	*upe* n. 'neck, nape of the neck; fig.: conscience' *lozo* vtr. 'to take, hold, get receive, affect sb.'	K, 493; F&S, 362, 312
4	*[dâŋ fe-fec] e-* Rdp-overshout do-	'to overshout each other'	*dâŋ* n. 'voice' *fezo* vintr. 'to go up, climb up'	K, 90; F&S, 84-85
5	*[feŋ(ko) qa-qac] e-* Rdp-dispute do-	'to quarrel, to dispute'	—	K, 90
6	*[afec ba-bac] e-* Rdp-brag do-	'to brag before each other'	*afecticne bazo* 'to praise sb.'	K, 2
7	*[micko ba-bac] e-* Rdp-dispute do-	'to dispute with each other'	*micko bazo* 'to answer him, to gainsay him'	K, 272
8	*[pelec e-ec] e-* Rdp-scold do-	'to scold one another'	*pele ezo* vintr. 'to scold'	P, 99; K, 330
9	*[tutuc qa-qac] e-* Rdp-reproach do-	'to accuse one another'	*tutuc qazo* 'to reproach, to accuse sb.'	K, 486
10	*[ʒâlâ ba-bac] e-* Rdp-quarrel do-	'to quarrel (with each other)'	*ʒâlâ bacnuzo* 'to prick, to scratch me; ie. to scold me violently'	K, 518
11	*[dâŋ mu-muc]* e-ec e- Rdp-scold Rdp-do do-	'to scold, abuse each other'	*dâŋ muzo* 'to scold, abuse sb.'	P, 99; K, 55
12	*[baŋko ba-bac] e-* Rdp-restrain do-	'to restrain each other from fighting'	*baŋko bazo* vtr. 'ward off fighting, restrain them, to help settle a dispute'	F&S, 25
13	*[soŋaŋ ba-bac] e-* Rdp-take.care do-	'to control each other, to have an eye upon each other'	*soŋaŋ bazo* 'to take care of, to mind sb. or sth.'	K, 435

(*Continued*)

Table 11.2.2 (Continued)

No.	*Reciprocal predicate*	*Interpretation*	*Base (<)*	*Source*
14	*[dâŋ ba-bac] e-* Rdp-talk do-	'to talk to each other'	*dâŋ ezo* 'to talk'	K, 54
15	*[mu qa-qazuc] e-e-* Rdp-teach Rdp-do-	'to teach one another'	*mu qazuc ezo* 'to teach, indoctrinate'	F&S, 223
16	*[ofac he-hec] e-* Rdp-call.oo do-	'to hail each other with calls of oo'	*ofac hezo* 'to call oo (to welcome the arriving guests: do come!)'	K, 318
17	*[kuku ʒi-ʒiliŋ] e-* Rdp-call.for.help do-	'to call each other for help, assistance'	*kuku qazo* 'to utter this call *kuku!*, to cry for help (e.g., against enemies)'	K, 219
18	*[qaŋgec qa-qac] e-* Rdp-talk.over do-	'to talk sth. over, (to exchange conjectures with each other)'	*qaŋgec qazo* 'to conjecture, to suppose'	K, 363
19	*[saŋuc qa-qac] e-* Rdp-whisper do-	'to whisper to each other'	*saŋuc qazo* 'to whisper'	K, 402
20	*[maŋjauŋ e-ec] e-* Rdp-love do-	'to love one another'	*maŋjauŋ ecnezo* 'to show him love'	F&S, 195
21	*[maŋ lâ-lâec] e-* Rdp-fall.in.love do-	'to fall in love with each other'	*maŋ lâezo* 'to fall in love; to set one's heart on sth.'	K, 260
22	*[tembeo lo-loc] e-* Rdp-seize.swing do-	1. 'to seize the swinging-rope and join the others in swinging' 2. 'to fall in love with each other, to be attached to each other'	*tembeo lozo* 'to seize the end of the rattan-swing in which sb. is swinging'	K, 468-69
23	*[sasec e-ec] e-* Rdp-lust do-	'to be passionate about each other'	*sasec ecnezo* 'to lust after sb.'	P, 99; F&S, 322
24	*[iiŋguuŋ e-ec] e-* Rdp-envy do-	'to envy each other'	*iiŋguuŋ ezo* 'to envy sb.'	K, 170
25	*[ʒaŋ ba-bac] e-* Rdp-be.afraid do-	'not to know each other, to be afraid of each other'	*ʒaŋ bazo* 'to be afraid, not to know (e.g., sb.'s character)'	K, 522
26	*[burec e-ec] e-* Rdp-lie do-	'to lie to one another'	*burec ezo* vintr. 'to lie, deceive'	P, 99; F&S, 55
27	*[sâc ba-bac] e-* Rdp-measure do-	'to compare one another'	*sâc bazo* vtr. 'to measure or weigh sth.'	F&S, 323
28	*[huc nâ-nâc] e-* Rdp-chew.betel do-	'to chew betel with each other, as sign for abolished hostility and restored peace'	*huc nâzo* 'to chew betel (things necessary for this: betel nut, betel pepper and lime)	K, 163

Table 11.2.2

No.	Reciprocal predicate	Interpretation	Base (<)	Source
29	*[uŋ ba-bac] e-* Rdp-caution do-	'to caution (warn) each other'	*uŋ bazo* vtr. 'to warn sb., to caution sb.'	K, 492; F&S, 362
30	*[jauŋ ba-bac] e-* Rdp-be.friendly do-	'to be kind and friendly to each other'	*jauŋ bazo* 'to be kind, bene-volent, friendly, good to sb.'	K, 187; F&S, 159
31	*[bu ririkec e-ec] e-* Rdp-incite do-	1. 'to animate/support each other' 2. 'to provoke one another'	*bu ririke ecnezo* vtr. 'to animate him, incite him, provoke him'	K, 234; F&S, 55
32	*[maŋzi ʒâ-ʒâfic] e-* Rdp-be.friend do-	'to establish fellowship'	*maŋzi ʒâfickezo* 'to be a true friend, to be united in love'	F&S, 200
33	*[due ba-bac] e-* Rdp-sacrifice do-	'to give one's lives for each other'	*due bazo* 'to sacrifice one's life for sb.'	K, 66-67
34	*[meo ba-bac] e-* Rdp-shake.hands do-	1. 'to shake hands' 2. 'to get married'	*meo bazo* 'shake hands, to take sb. by the hand'	K, 31; F&S, 207
35	*[meo lo-loc] e-* Rdp-unite do-	'to unite, to take one another by the hand'	*meo lozo* 'to take by the hands, shake hands, to unite'	F&S, 207
36	*[boe ba-bac] e-* Rdp-adultery do-	'to be in illicit sexual intercourse'	*boe* n. 'debauchee, prostitute, slut'	K, 44
37	*[sau ba-bac] e-* Rdp-adultery do-	'to have intercourse with each other'	*sau bazo* 'to entertain adulterous relations, ie. to commit adultery, to live dissolutely'	K, 406
38	*[mu simbiŋ e-ec] e-* Rdp-seduce do- *mu-musimbiŋ e-* Rdp-seduce do-	'to seduce one another'	*mu simbiŋ ezo* 'to convince sb., to seduce sb., to take part in a dirty deal'	F&S, 225; K, 289
39	*[boleŋ ba-bac] e-* Rdp-exchange. do do-	'to exchange ornaments in token reconciliation'	*boleŋ* n. 'valuable orna-ments like boar's tusks, pockets adorned with dog's fangs, head-band made of dog's fangs etc.'	K, 46
40	*[ipuc ba-bac] e-* Rdp-exchange. do do-	'to exchange expiatory gifts'	*ipu, ipune* n. 'expiatory gift for men slain on both sides (at the conclusion of peace a price must be paid for every person that has been slain)	K, 176; F&S, 154

(*Continued*)

Table 11.2.2 (Continued)

No.	Reciprocal predicate	Interpretation	Base (<)	Source
41	*[amao he-hec] e-* Rdp-encounter do-	1. 'to bump into one another (in the darkness)' 2. 'to encounter each other (hostile divisions)'	—	K, 7; F&S, 4
42	*[âgo ba-bac] e-* Rdp-make.friends do-	'to make/become friends with one another'	*âgo bazo* vtr. 'to make friends with sb.'	K, 3; F&S, 9

Source: Author created.

Table 11.2.3 Compositional Reciprocal Verbs in Kâte (Keyßer, 1969 [1925]; Pilhofer, 1969 [1933]; Flierl & Strauss, 1977).

No.	Reciprocal predicate	Interpretation	Base (<)	Source
1	*nowac(loŋ) gowacloŋ e-* take.away.1Sg/2Sg do-	'to take sth. away from each other'	*nowaclozo* 'to take sth. away from me'	K, 303; P, 99
2	*nowac(nâŋ) gowacnâŋ e-* eat.from.1Sg/2Sg do-	'to eat each other's food'	*nowacnâzo* 'to eat of my portion'	P, 99; K, 303
3	*nazaŋ gazaŋ e-* speak.1Sg/2Sg do-	'to speak to one another, to converse'	*nazazo* 'to tell me sth.'	P, 99; K, 297
4	*[uŋ nazaŋgazaŋ] e-* caution.1Sg/2Sg do-	'to caution (warn) each other'	*nazaŋ gazaŋ e-* 'speak to one another' < *nazazo* 'to tell me sth.'	K, 492; F&S, 362
5	*jowac gowac e-* lead.3Sg/2Sg do-	'to lead each other, to walk together'	*jowazo* 'to lead sb. or sth.'	K, 190
6	*naleŋ galeŋ e-* give.1Sg/2Sg do-	'to give each other, to exchange, to barter'	*nalezo* 'to give me sth.'	P, 99; K, 293
7	*nowac(luŋ) gowaluŋ e-* go.past.1Sg/2Sg do-	'to go past each other'	*nowaluzo* 'to go past me'	K, 303; P, 99
8	*napeŋ gapeŋ e-* follow.1Sg/2Sg do-	'to follow each other, to walk behind one another'	*napezo* 'to walk behind me, to follow me'	P, 99; K, 296

Source: Author created.

NYULNYULAN LANGUAGES

The tables below, Tables 11.3.1–11.3.2, are complementary to Figure 5.3 in chapter 5.

Table 11.3.1 Reciprocal Use of Verbal Circumfix *ma-V-nyji* in Nyulnyulan Languages (McGregor, 1999).

No.	*Reciprocal form*	*Interpretation*	*Base (<)*	*Source*
1	*-ma-yama-nyji-* -RR-argue-RR-	'to argue with each other'	*-yama-* vtr. 'to argue'	McG, 89
2	*-ma-kama-nyji-* -RR-mock-RR-	'to mock one another'	*-kama-* vtr. 'to mock'	McG, 90
3	*-ma-bula-nyji-* -RR-meet-RR-	'to meet together'	*-bula-* vintr. 'to come'	McG, 104
4	*-mi-jal-inyj-* -RR-see-RR-	'to look at each other'	*-jal-* vtr. 'to look at'	McG, 91
5	*-mi-j-inyj-* -RR-say-RR-	'to talk to one another'	*-j-* vtr. 'to say'	McG, 94
6	*-ma-wara-nyji-* -RR-follow-RR-	'to follow each other'	*-wara-* vtr. 'to follow'	McG, 90
7	*-ma-ng-ka-nyji-* -RR-E:Nasal-hit-RR-	'to hit/fight each other'	*-ka-* vtr. 'to hit'	McG, 89

Source: Author created.

Table 11.3.2 Reciprocal Use of Auxiliary *-banji/-wanji* in Nyulnyulan Languages (McGregor, 1999).

No.	*Reciprocal form*	*Interpretation*	*Base (<)*	*Source*
1	*ngaw -banji-* kiss -RR-	'to kiss one another'	*ngaw* 'kiss'	McG, 96
2	*kanyjirr -wanji-* look -RR-	'to stare at one another'	*kanyjirr* 'look'	McG, 97
3	*ngul -wanji-* spear -RR-	'to spear one another'	*ngul* 'spear'	McG, 107
4	*mad -banj-* shake -RR-	'to shake hands'	*mad* 'shake'	McG, 97
5	*wilamay -banj* food -RR-	'to give each other food'	(*wilamay* 'food')	McG, 95

Source: Author created.

KU WARU

The following table, Table 11.4, is complementary to Figure 5.4 in chapter 5.

Table 11.4 Reciprocal Use of Compositional Verb Forms in Ku Waru (Merlan & Rumsey, 1991, Appendix A, and Merlan & Rumsey, 1989).

No.	*Predicate*	*Interpretation*	*Base (<)*	*Source*
1	*[el ung] nyi-* fight.talk speak-	'to speak *el ung* with each other'	*el ung nyi-* n^v 'to speak fight talk'?	M&R, 1991, line 33
2	*el te-* fight do-	'to fight with each other about something'	*el te-* n^v 'to fight with'?	M&R, 1991, line 90
3	*el to-* fight hit-	'to fight against each other'	*el to-* n^v 'to fight against'?	M&R, 1991, line 453
4	*[pula to-] abul-* break- hold-	'to fight each other'	(*pula to-* p^v 'to break/cut') (*abul-* v 'to hold')	M&R, 1991, line 415; M&R, 1989, 17, 1706 .kw
5	*[kapu el] te-* fight do-	'to fight over (things)'	*kapu el te-* 'to fight with'? (*kapu el* n^n 'fight arising from exchange')	M&R, 1991, line 1157; M&R, 1989, 474 .kw
6	*nyi- te-* talk- do-	'to say to each other'	*nyi- te-* v^v 'to say to'?	M&R, 1991, line 441
7	*nyi-* talk-	'to talk to each other about sth.'	*nyi-* v 'to talk (about)'?	M&R, 1991, line 363
8	*para nyi-* level talk-	'to talk back and forth'	(*nyi-* v 'to talk') (*para* p 'back and forth')	M&R, 1991, line 1094
9	*tobul- nyi-* discuss- say-	'to discuss sth. with each other'	(*tobul-* v 'discuss') (*nyi-* v 'talk')	M&R, 1991, line 189
10	*ung [dupi- si-]* talk tell.story-	'telling each other stories'	*dupi- si-* v 'to tell story'	M&R, 1989, 219 .kw
11	*kudi to-* tell.secrets hit-	'to rat on each other (e.g., children)'	*kudi* n 'tell secrets; rat'	M&R, 1989, 805 .kw
12	*kana-* see-	'to see each other'	*kana-* v 'to look at, see'	M&R, 1989, 447, 2137 .kw
13	*maku to-* bundle hit-	'to meet, come together, gather'	*maku to-* 'to meet with'?	M&R, 1991, line 363; M&R, 1989, 1106-7
14	*miting te-* meeting do-	'to have a meeting together'	*miting te-* 'to have a meeting with'?	M&R, 1991, line 361

Table 11.4

No.	*Predicate*	*Interpretation*	*Base (<)*	*Source*
15	*tada-* join-	'to meet up, join together (e.g., rivers)'	*tada-* v 'to join, meet up'	M&R, 1989, 1880 .kw
16	*o-* come-	'to assemble, come together'	*o-* v 'to come'	M&R, 1991, line 1436
17	*lupu lupu pu-* other other go-	'to disperse, go in different directions'	(*lupu* 'other, different') (*pu-* v 'to go')	M&R, 1991, line 1596, line 631
18	*elkupula te-* push/shove-	'elbow and shove each other'	*elkupula te-* p^v 'to push and shove; to elbow each other'	M&R, 1989, 247 .kw
19	*elkelipa pu-* choke-	'to push each other'	*elkelipa pu-* p^v 'choke, block, crowd; to push each other'	M&R, 1989, 242 .kw
20	*lyi- kumulau* clutch-	'clutching each other'	*kumulau lyi-* p^v 'to clutch, grip'	M&R, 1989, 870 .kw, 1598 .kw
21	*opimi to-* join-	'to join together (e.g., in fighting)'	*opimi to-* p^v 'to add, contribute; join together, become friendly'	M&R, 1989, 1469 .kw; M&R, 1991, line 529
22	*ung [opimi to-]* word join-	'make promises to each other'	*ung opimi to-* n^[p^v] 'to promise'	M&R, 1989, 1469 .kw
23	*wai to-* make.promise-	'to make arrangements'	*wai to-* p^v 'to promise, make an arrangement with'	M&R, 1989, 2137 .kw
24	*orara te-* cooperate-	'to cooperate with each other'	*orara te-* p^v 'to cooperate, work together; '	M&R, 1989, 1474 .kw
25	*helpim te-* help-	'to help each other'	*helpim te-* 'to help sb.'?	M&R, 1989, 1474 .kw
26	*kapu te- ti-* fight do- give-	'to fight with each other (because of an unmet expectation that one will give to the other but hasn't done so)	*kapu te-* n^v 'to fight over things' *ti-* v 'to give'	M&R, 1991, line 1155 (& Rumsey, p.c.)
27	*orim to-* share-	'to share, divide'***	—	M&R, 1989, 1475 .kw

(*Continued*)

Table 11.4 (Continued)

No.	*Predicate*	*Interpretation*	*Base (<)*	*Source*
28	*kolum te-* cut-	'to break up, divide'***	*kolum te-* p^v 'to cut'	M&R, 1991, line 1318, M&R, 1989, 720 .kw
29	*warltowarlto te-* Rdp.small do-	'to share, divide'***	(*warlto* n 'half, small, little, little bit')	M&R, 1989, 2195 .kw
30	*kelkel te-* divide-	'to share, divide (land)'***	—	M&R, 1989, 534 .kw
31	*oi te-* demarcate-	'to divide (land)'***	*oi te-* n^v 'to make a demarcation' (*oi* n 'border, boundary')	M&R, 1989, 1428 .kw

***These various forms with meanings like 'share, divide' are included in the list of possible reciprocals here as (1) they at least partly fall under the a priori definition of reciprocity as implying "mutual" or "symmetrical" relationships (i.e., they are inherently reciprocal), (2) because in other languages (e.g., Nungubuyu, Yidiny), explicit reciprocal marking is in some cases used to encode such meanings, and (3) since the question of a possible close link of reciprocal marking to the expression of concepts of social reciprocity requires the inclusion of such notions into the present discussion (see chapter 8).
Source: Author created.

Appendix C

References for Figure Entries

Table 12.1 References for Entries Made in Figures 3.1 and 3.2, resp. (chapter 3). Underlinings indicate shadings in the original figures.

	Known to be available (see reference after each entry)	*Known to be unavailable (see reference after each entry)*	*Not known whether available or unavailable*
Bossou	*Elaeis guineensis* (Biro, 2011, pp. 166-167) *Parinari excelsa* (ibid.) *Treculia africana* (Koops, 2010, p. 177)	*Coula sp.* (Biro, 2011, pp. 166-67) *Sacoglottis sp.* (ibid.) *Detarium sp.* (ibid.) *Strychnos sp.* (ibid.) *Panda sp.* (ibid.) *Carapa sp.* (ibid.) *Maranthes sp.* (ibid.) *Klainedoxa sp.* (ibid.)	*Achantina sp.*
Taï	*Elaeis guineensis* (Boesch et al., 1994, pp. 328-29) *Coula edulis* (ibid.) *Deterium senegalensis* (ibid.) *Parinari excelsa* (ibid.) *Panda oleosa* (ibid.) *Sacoglottis gabonensis* (ibid.) *Strychnos sp.* (Whiten et al., 2001, p. 1503) *Treculia africana* (Koops et al., 2010, p. 177)		*Carapa sp.* *Achantina sp.* *Maranthes sp.* *Klainedoxa sp.*

(*Continued*)

Table 12.1 (Continued)

	Known to be available (see reference after each entry)	*Known to be unavailable (see reference after each entry)*	*Not known whether available or unavailable*
Gombe	*Elaeis guineensis* (Sugiyama, 1985, p. 372) *Strychnos sp.* (ibid.)		*Coula sp.* *Detarium sp.* *Parinari sp.* *Panda sp.* *Sacoglottis sp.* *Carapa sp.* *Achantina sp.* *Maranthes sp.* *Klainedoxa sp.* *Treculia sp.*
Lopé	*Elaeis guineensis* (McGrew et al., 1997, p. 358) *Coula edulis* (ibid.) *Detarium macrocarpum* (ibid.) *Panda oleosa* (ibid.) *Sacoglottis gabonensis* (ibid.) *Maranthes glabra* (ibid.) *Klainedoxa gabonensis* (ibid.)		*Strychnos sp.* *Treculia sp.* *Parinari sp.* *Carapa sp.* *Achantina sp.*
Yealé	*Elaeis guineensis* (Humle, 2011b, p. 270) *Coula edulis* (ibid.) *Parinari excelsa* (ibid.) *Panda oleosa* (ibid.) *Detarium senegalensis* (ibid.) *Strychnos sp.* (ibid.) *Carapa procera* (ibid.) *Achantina sp.* (ibid.) *Treculia africana* (Koops, 2010, p. 177)		*Sacoglottis sp.* *Maranthes sp.* *Klainedoxa sp.*
Seringbara	*Elaeis guineensis* (Koops, 2011, p. 280) *Detarium senegalensis* (ibid.) *Parinari excelsa* (ibid.) *Parinari glabra* (ibid.) *Treculia africana* (Koops, 2010)	*Coula edulis* (Biro, 2011, p. 167) *Panda oleosa* (ibid.)	*Sacoglottis sp.* *Strychnos sp.* *Carapa sp.* *Achantina sp.* *Maranthes sp.* *Klainedoxa sp.*

Table 12.1

	Known to be available (see reference after each entry)	*Known to be unavailable (see reference after each entry)*	*Not known whether available or unavailable*
Diécké	*Elaeis guineensis* (Biro, 2011, p. 167) *Coula edulis* (Carvalho, 2011, p. 304) *Parinari excelsa* (ibid.) *Panda oleosa* (ibid.) *Strychnos spinosa* (ibid.)		*Detarium sp.* *Sacoglottis sp.* *Treculia africana* *Carapa sp.* *Achantina sp.* *Maranthes sp.* *Klainedoxa sp.*

Source: Author created.

Table 12.2 References for Entries in Figure 6.1 (chapter 6).

No.	*Yidiny*	*Dyirbal*	*Dyabugay*
1	Dixon, 1977: 380, 370	Dixon, 1972: 93	Patz, 1991: 296
2	Dixon, 1977: 380	Dixon, 1972: 388, 398	Patz, 1991: 283
3	Dixon, 1977: 380	Dixon, 1972: 93	Patz, 1991: 296
4	Dixon, 1977: 380, 470	Dixon, 1972: 92	Patz, 1991: 297
5	Dixon, 1977: 379-380	Dixon, 1972: 93	Patz, 1991: 296
6	Dixon, 1977: 282	Dixon, 1972: 299	—
7	Dixon, 1977: 162	Dixon, 1972: 300	—
8	—	—	Patz, 1991: 283
9	—	—	Patz, 1991: 297

Source: Author created.

Table 12.3 References for Entries in Figure 6.2 (chapter 6).

No.	*Jiwarli*	*Martuthunira*	*Yindjibarndi*	*Panyjima*
1	—	Dench, 1995: T5.1-2	—	—
2	—	Dench, 1995: T4.54	—	—
3	—	Dench, 1995: 69	—	—
4	—	Dench, 1995: 107	—	—
5	—	Dench, 1995: 107	—	Dench, 1987: 330
6	Burgman, 2006: 18; Austin, 1992: 44	Dench, 1995: 4.141, p. 92	Wordick, 1982: 91	Dench, 1991: 191
7	Burgman, 2006: 16; Austin, 1992: 40	Dench, 1995: T7.338, p. 311	Wordick, 1982: T60.4, p. 235	Dench, 1991: 191
8	Burgman, 2006: 20; Austin, 1992: 49	Dench, 1995: 6.46, p. 153	Wordick, 1982: T66.8, p. 243	Dench, 1991: 172
9	Burgman, 2006: 21; Austin, 1992: 52	Dench, 1995: T4.41, p. 277	Wordick, 1982: 91	Dench, 1991: 147
10	P. Austin, elicited material, p.c., 2010	Dench, 1995: 6.46, p. 153	Wordick, 1982: 90	Dench, 1991: 152
11	Austin, 1992: 60	Dench, 1995: T7.330, p. 311	Wordick, 1982: 90	Dench, 1991: 202
12	Burgman, 2006: 24; Austin, 1992: 58	Dench, 1995: T4.24, p. 276	Wordick, 1982: T60.5, p. 235	Dench, 1991: T6.4, line 17, p. 226
13	P. Austin, elicited material, p.c., 2010	Dench, 1995: 4.133, p. 91	Wordick, 1982: 90	Dench, 1991: 191
14	Burgman, 2006: 20 Austin, 1992: 49	Dench, 1995: 6.48, p. 153	Wordick, 1982: 90	Dench, 1991: 192
15	Burgman, 2006: 13; Austin, 1992: 29	Dench, 1995: T1.33, p. 266	Wordick, 1982: 90	Dench, 1991: 192
16	P. Austin, elicited material, p.c., 2010	Dench, 1995: T7.131, p. 296	—	Dench, 1991: 192
17	—	—	—	Dench, 1991: 192

Source: Author created.

Table 12.4 References for Entries in Figure 6.3 (chapter 6).

No.	*Nunggubuyu*	*Nyulnyulan*
1	Heath, 1980: T73.3, p. 334	McGregor, 1999: 89
2	Heath, 1980: T69.13, p. 313	Stokes, 1982: 289
3	Heath, 1980: T69.13, p. 313	Stokes, 1982: 290
4	Heath, 1980: T75.1, p. 339	McGregor, 1999: 91
5	Heath, 1980: T76.2, p. 340	Nekes & Worms, 1953: 137
6	Heath, 1980: T46.1, p. 243	McGregor, 1999: 90
7	Heath, 1980: T5.20, p. 46	McGregor, 1999: 89
8	Heath, 1980: T7.8, p. 52	McGregor, 1999: 89
9	Heath, 1981: 168	McGregor, 1999: 89
10	Heath, 1981: 150	Nekes & Worms, 1953: 385
11	Heath, 1984: 390	McGregor, 1999: 91
12	Heath, 1984: 390	McGregor, 1999: 92
13	Heath, 1984: 390	McGregor, 1999: 100
14	Heath, 1981: 193	Stokes, 1982: 289

Source: Author created.

Table 12.5 References for Entries in Table 8.1 (chapter 8). The information given in brackets (right-hand column 'Table Entry') refer to Appendix B, Tables 11.2.1 and 11.2.2.

Concept	*Reciprocal verb*	*Table Entry*
'exchange'	*boleŋ babac ezo*	Keyßer, 1969: 46 (app. B, tab. 11.2.2, no. 39)
'make peace'	*huc nânâc ezo*	Keyßer, 1969: 163 (app. B, tab. 11.2.2, no. 28)
'assist; unite'	*ʒiʒimic ezo*	Keyßer, 1969: 537-38 (app. B, tab. 11.2.1, no. 39)
'help'	*ŋiŋiec ezo*	Keyßer, 1969: 313 (app. B, tab. 11.2.1, no. 38)
'make war'	*qaqac ezo*	Keyßer, 1969: 373; Flierl & Strauss, 1977: 279 (app. B, tab. 11.2.1, no. 1)
'take revenge'	*okac babac ezo*	Flierl & Strauss, 1977: 253; Pilhofer, 1969: 99; Keyßer, 1969: 319 (app. B, tab. 11.2.2, no. 1)
'outdo'	*fefec ezo*	Keyßer, 1969: 90; Flierl & Strauss, 1977: 84-85 (App. B, tab. 11.2.1, no. 27)

Source: Author created.

References

Aikhenvald, A. Y. (2013). "Possession and ownership: A cross-linguistic perspective." In A. Y. Aikhenvald & R. M. W. Dixon (Eds.), *Possession and ownership: A cross-linguistic typology* (pp. 1–64). Oxford: Oxford University Press.

Aissen, J. (2003). "Differential object marking: Iconicity vs. economy." *Natural Language and Linguistic Theory, 21*(3), 435–483.

Alpher, B., Evans, N., & Harvey, M. (2003). "Proto-Gunwinyguan verb suffixes." In N. Evans (Ed.), *The non-Pama-Nyungan languages of Northern Australia: Comparative studies of the continent's most linguistically complex region* (pp. 305–352). Canberra: Pacific Linguistics.

Arthur, W. B. (2007). "The structure of invention." *Research Policy, 36,* 274–287.

Austin, P. (1992). *A dictionary of Jiwarli, Western Australia.* La Trobe University, Department of Linguistics.

———. (1997). *Texts in the Mantharta languages, Western Australia.* ILCAA, Tokyo University of Foreign Studies.

Axelrod, R. (1984). *The evolution of cooperation.* New York: Basic Books.

Barker, C. (1995). *Possessive descriptions.* Stanford, CA: CSLI Publications.

Bauer, L. (2001). *Morphological productivity.* Cambridge: Cambridge University Press.

Beck, B. B. (1980). *Animal tool behavior.* New York: Garland Press.

Belletti, A., & Rizzi, L. (Eds.) (2002). *On nature and language.* Cambridge: Cambridge University Press.

Benveniste, E. (1971 [1939]). "The nature of the linguistic sign." In E. Benveniste (Ed.), *Problems in general linguistics* (pp. 43–48). Translated by M. Meek. Coral Gables, FL: University of Miami Press.

Berg, J., Dickhaut, J., McCabe, K. (1995). "Trust, reciprocity, and social history." *Games and Economic Behavior, 10*(1), 122–142.

Berwick, R. C., & Chomsky, N. (2016). *Why only us? Language and evolution.* Cambridge, MA: The MIT Press.

Bickel, B. (2003). "Referential density in discourse and syntactic typology." *Language, 79*(4), 708–736.

Bickerton, D. (1996). *Language and human behavior.* London: University College London Press.

Biro, D. (2011). "Clues to culture? The Coula- and Panda-nut experiments." In T. Matsuzawa, T. Humle, & Y. Sugiyama (Eds.), *The chimpanzees of Bossou and Nimba* (pp. 165–173). Tokyo: Springer.

Boeckx, C., & Piattelli-Palmarini, M. (2005). "Language as a natural object, linguistics as a natural science." *The Linguistic Review, 22,* 447–466.

Boesch, C. (2002). "Cooperative hunting roles among Taï chimpanzees." *Human Nature, 13*(1), 27–46.

———. (2012). "From material to symbolic cultures: Culture in primates." In J. Valsiner (Ed.), *The oxford handbook of culture and psychology* (pp. 677–694). Oxford: Oxford University Press.

———. (2013). "Ecology and cognition of tool use in chimpanzees." In C. M. Sanz, J. Call, & C. Boesch (Eds.), *Tool use in animals: Cognition and ecology* (pp. 21–47). Cambridge: Cambridge University Press.

Boesch, C., & Boesch, H. (1983). "Optimization of nut-cracking with natural hammers by wild chimpanzees." *Behaviour, 83,* 265–286.

———. (1990). "Tool use and tool making in wild chimpanzees." *Folia Primatologica, 54,* 86–99.

Boesch, C., Marchesi, P., Marchesi, N., Fruth, B., & Joulian, F. (1994). "Is nut cracking in wild chimpanzees a cultural behavior?" *Journal of Human Evolution, 26,* 325–338.

Bolinger, D. L. (1978). "Intonation across languages." In J. P. Greenberg, C. A. Ferguson, & E. A. Moravcsik (Eds.), *Universals of human language, volume 2: Phonology* (pp. 471–525). Palo Alto, CA: Stanford University Press.

Bonnie, K. E., Horner, V., Whiten, A., & De Waal, F. B. M. (2007). "Spread of arbitrary customs among chimpanzees: A controlled experiment." *Proceedings of the Royal Society B, 274,* 367–372.

Boroditsky, L. (2001). "Does language shape thought? English and Mandarin speakers' conceptions of time." *Cognitive Psychology, 43*(1), 1–22.

Boroditsky, L., & Gaby, A. (2010). "Remembrances of times east: Absolute spatial representations of time in an Australian Aboriginal community." *Psychological Science, 21*(11), 1635–1639.

Brentano, F. (1973 [1874]). *Psychology from an empirical standpoint.* Translated by A. C. Rancurello, D. B. Terrell, & L. McAlister. London: Routledge.

Brownie, J., & Brownie, M. (2007). *Massau grammar essentials.* Ukarumpa: SIL-PNG Academic Publications.

Bühler, K. (1934). *Sprachtheorie: Die Darstellungsfunktion der Sprache* [Theory of language: The representational function of language]. Jena: Fischer.

Burenhult, N. (2011). "The coding of reciprocal events in Jahai." In N. Evans, A. Gaby, S. C. Levinson, & A. Majid (Eds.), *Reciprocals and semantic typology* (pp. 163–176). Amsterdam: John Benjamins.

Burgman, A. (2006). *Draft Jiwarli dictionary and topical finder lists 2006.* South Hedland, WA: Wangka Maya Pilbara Aboriginal Language Centre.

Burling, R. (2005). *The talking ape: How language evolved.* Oxford: Oxford University Press.

Bybee, J. L., Perkins, R. D., & Pagliuca, W. (1994). *The evolution of grammar: Tense, aspect, and modality in the languages of the world.* Chicago: University of Chicago Press.

Byrne, R. W., (1995). *The thinking ape: Evolutionary origins of intelligence.* Oxford: Oxford University Press.

Byrne, R. W., & Russon, A. E. (1998). "Learning by imitation: A hierarchical approach." *Behavioral and Brain Sciences, 21,* 667–684.

Call, J. (2013). "Three ingredients for becoming a creative tool user." In C. M. Sanz, J. Call, & C. Boesch (Eds.), *Tool use in animals: Cognition and ecology* (pp. 3–20). Cambridge: Cambridge University Press.

Call, J., & Tomasello, M. (2007). "Comparing the gestural repertoire of apes." In J. Call & M. Tomasello (Eds.), *The gestural communication of apes and monkeys* (pp. 197–220). New York: LEA.

Campbell, J. (2011). "Why do language use and tool use both count as manifestations of intelligence?" In T. McCormack, C. Hoerl, & S. Butterfill (Eds.), *Tool use and causal cognition* (pp. 169–182). Oxford: Oxford University Press.

Carew, T. J. (2000). *Behavioral neurobiology: The cellular organization of natural behavior.* Sunderland, MA: Sinauer Associates.

Carvalho, S. (2011). "Diécké Forest, Guinea: Delving into chimpanzee behavior using stone tool surveys." In T. Matsuzawa, T. Humle, & Y. Sugiyama (Eds.), *The chimpanzees of Bossou and Nimba* (pp. 301–312). Tokyo: Springer.

Chafe, W. (1980). *The pear stories: Cognitive, cultural, and linguistic aspects of narrative production.* Norwood, NJ: Ablex.

Chan, W. P., Prete, F., & Dickinson, M. (1998). "Visual input to the efferent control system of a fly's 'gyroscope.'" *Science, 280,* 289–292.

Chappell, H., & McGregor, W. B. (1995). "Prolegomena to a theory of inalienability." In H. Chappell & W. B. McGregor (Eds.), *The grammar of inalienability: A typological perspective on body part terms and the part-whole relation* (pp. 3–30). Berlin: Mouton de Gruyter.

Cheour, M., Ceponiene, R., Lehtokoski, A., Luuk, A., Allik, J., Alho, K., & Näätänen, R. (1998). "Development of language-specific phoneme representations in the infant brain." *Nature Neuroscience, 1*(5), 351–353.

Chomsky, N. (1959). "A review of Skinner's Verbal Behavior." *Language, 35*(1), 26–58.

———. (1965). *Aspects of the theory of syntax.* Cambridge, MA: The MIT Press.

———. (1975). *Reflections on language.* New York: Pantheon Books.

———. (1980). *Rules and representations.* New York: Columbia University Press.

———. (1981). *Lectures on government and binding.* Holland: Foris Publications.

———. (1995). *The minimalist program.* Cambridge, MA: The MIT Press.

———. (2000). *New horizons in the study of language and mind.* Cambridge: Cambridge University Press.

———. (2002a). "Language and the brain." In A. Belletti & L. Rizzi (Eds.), *On nature and language.* Cambridge: Cambridge University Press.

———. (2002b). "An interview on minimalism." In A. Belletti & L. Rizzi (Eds.), *On nature and language*. Cambridge: Cambridge University Press.

———. (2009). "Opening remarks." In M. Piattelli-Palmarini, J. Uriagereka, & P. Salaburu (Eds.), *Of minds and language: A dialogue with Noam Chomsky in the Basque country* (pp. 13–43). Oxford: Oxford University Press.

———. (2010). "Some simple evo devo theses: How true might they be for language?" In R. K. Larson, V. Déprez, & H. Yamakido (Eds.), *The evolution of human language: Biolinguistic perspectives* (pp. 45–62). Cambridge: Cambridge University Press.

Christiansen, M. H., & Chater, N. (2008). "Language as shaped by the brain." *Behavioral and Brain Sciences, 31*, 489–558.

Clayton, N. S., & Dickinson, A. (1998). "Episodic-like memory during cache recovery by scrub jays." *Nature, 395,* 272–274.

Clayton, N. S., Griffiths, D., Emery, N., & Dickinson, A. (2001). "Elements of episodic-like memory in animals." *Philosophical Transactions of the Royal Society of London B, 356*(1413), 1483–1491.

Comrie, B. (1989). *Language universals and linguistic typology,* second edition. Oxford: Blackwell.

———. (1998). "Rethinking the typology of relative clauses." *Language Design, 1,* 59–86.

Cook, K. S., Cheshire, C., Rice, E. R. W., & Nakagawa, S. (2013). "Social exchange theory." In J. DeLamater & A. Ward (Eds.), *Handbook of social psychology* (pp. 61–88). Dordrecht: Springer.

Corballis, M. C. (2011). *The recursive mind: The origins of human language, thought, and civilization.* Princeton, NJ: Princeton University Press.

Corbett, G. G., & Noonan, M. (Eds.) (2008). *Case and grammatical relations. Studies in honor of Bernard Comrie.* Amsterdam: John Benjamins.

Cowie, F. (2017 [2008]). "Innateness and language." In E. N. Zalta (Ed.), *The Stanford Encyclopedia of Philosophy*, fall 2017 edition. https://plato.stanford.edu/archives/fall2017/entries/innateness-language/.

Cowley, S. J. (Ed.) (2011a). *Distributed language.* Amsterdam: John Benjamins.

———. (2011b). "Distributed language." In S. J. Cowley (Ed.), *Distributed language* (pp. 1–14). Amsterdam: John Benjamins.

———. (2016). "Changing the idea of language: Nigel Love's perspective." *Language Sciences, 61,* 43–55.

Cowley, S. J., & Vallée-Tourangeau, F. (2017 [2013]). "Thinking, values and meaning in changing cognitive ecologies." In S. J. Cowley & F. Vallée-Tourangeau (Eds.), *Cognition beyond the brain: Computation, interactivity and human artifice,* second edition (pp. 1–17). London: Springer.

Croft, W. (2000). *Explaining language change: An evolutionary approach.* Harlow, Essex: Longman.

Danziger, E. (2008). "A person, a place, or a thing? Whorfian consequences of syntactic bootstrapping in Mopan Maya." In M. Bowerman & P. Brown (Eds.), *Crosslinguistic perspectives on argument structure* (pp. 29–48). New York: Erlbaum.

Darwin, C. 2004 [1879]. *The Descent of Man.* London: Penguin Books.

Davidson, J. E. (1995). "The suddenness of insight." In R. J. Sternberg & J. E. Davidson (Eds.), *The nature of insight* (pp. 125–55). Cambridge, MA: The MIT Press.

Deacon, T. W. (1997). *The symbolic species: The co-evolution of language and the brain.* New York: Norton.

Deary, I. J. (2001). *Intelligence: A very short introduction.* Oxford: Oxford University Press.

Dehaene, S. (2014). *Consciousness and the brain: Deciphering how the brain codes our thoughts.* New York: Penguin Books.

Dench, A. (1981). "Panyjima phonology and morphology." Master's thesis, Canberra, Australian National University.

———. (1982). "The development of an accusative case marking pattern in the Ngayarda languages of Western Australia." *Australian Journal of Linguistics, 2,* 43–59.

———. (1987). "Kinship and collective activity in the Ngayarda languages of Australia." *Language in Society, 16,* 321–340.

———. (1991). "Panyjima." In R. M. W. Dixon & B. J. Blake (Eds.), *The handbook of Australian languages.* Vol. 4 (pp. 125–243). Oxford: Oxford University Press.

———. (1995). *Martuthunira: A language of the Pilbara Region of Western Australia.* Canberra: Pacific Linguistics.

De Waal, F. B. M. (1997). "The chimpanzee's service economy: Food for grooming." *Evolution and Human Behavior, 18,* 375–386.

———. (2005a). *Our inner ape.* New York: Riverhead Books.

———. (2005b). "How animals do business." *Scientific American, 292*(4), 72–79.

Dimitriadis, A. (2008). "Irreducible symmetry in reciprocal constructions." In E. König & V. Gast (Eds.), *Reciprocals and reflexives: Theoretical and typological explorations* (pp. 375–409). Berlin: Mouton de Gruyter.

Dixon, R. M. W. (1972). *The Dyirbal language of North Queensland.* Cambridge: Cambridge University Press.

———. (1977). *A grammar of Yidiny.* Cambridge: Cambridge University Press.

———. (1980). *The languages of Australia.* Cambridge: Cambridge University Press.

———. (1988). *A grammar of Boumaa Fijian.* Chicago: University of Chicago Press.

———. (1991). *Words of our country: Stories, place names and vocabulary in Yidiny, the Aboriginal language of the Cairns-Yarrabah region.* Compiled and edited by R. M. W. Dixon. Brisbane: University of Queensland Press.

———. (2002). *Australian languages: Their nature and development.* Cambridge: Cambridge University Press.

Dixon, R. M. W., & Blake, B. J. (1991). *The handbook of Australian languages.* Vol. 4. Oxford: Oxford University Press.

Donald, M. (1991). *Origins of the modern mind: Three stages in the evolution of culture and cognition.* Cambridge, MA: Harvard University Press.

Donaldson, T. (1980). *Ngiyambaa: The language of the Wangaaybuwan.* Cambridge: Cambridge University Press.

Donohue, M. (2004). *A grammar of the Skou language of New Guinea.* University of Singapore. Online publication.

Dor, D. (2011). "Language as a communication technology: A proposal for a new general linguistic theory." http://people.socsci.tau.ac.il/mu/danield/linguistics-2/language-as-a-communication-technology/.

———. (2014). "The instruction of imagination: Language and its evolution as a communication technology." In D. Dor, C. Knight, & J. Lewis (Eds.), *The social origins of language* (pp. 105–125). Oxford: Oxford University Press.

Dor, D., Knight, C., & Lewis, J. (2014). "Introduction: A social perspective on how language began." In D. Dor, C. Knight, & J. Lewis (Eds.), *The social origins of language* (pp. 1–12). Oxford: Oxford University Press.

Duranti, A. (2011). "Linguistic anthropology: The study of language as a non-neutral medium." In R. Mesthrie (Ed.), *The Cambridge handbook of sociolinguistics* (pp. 28–46). Cambridge: Cambridge University Press.

Enfield, N. J. (2011). "Description of reciprocal situations in Lao." In N. Evans, A. Gaby, S. C. Levinson, & A. Majid (Eds.), *Reciprocals and semantic typology* (pp. 129–148). Amsterdam: John Benjamins.

Enfield, N. J., & Levinson, S. C. (2006). "Introduction: Human sociality as a new interdisciplinary field." In N. J. Emfield & S. C. Levinson (Eds.), *Roots of human sociality: Culture, cognition and interaction* (pp. 1–35). Oxford: Berg.

Engelmann, J. M., Herrmann, E., & Tomasello, M. (2015). "Chimpanzees trust conspecifics to engage in low-cost reciprocity." *Proceedings of the Royal Society B, 282,* 20142803.

Evans, N. (2007). "Reciprocals, grammar-writing and typology." *Revue de Linguistique Roumaine, II*(4), 479–506.

———. (2008). "Reciprocal constructions: Towards a structural typology." In E. König & V. Gast (Eds.), *Reciprocals and reflexives: Theoretical and typological explorations* (pp. 33–104). Berlin: Mouton de Gruyter.

Evans, N., & Levinson, S. C. (2009). "The myth of language universals: Language diversity and its importance for cognitive science." *Behavioral and Brain Sciences, 32,* 429–448.

Evans, N., Levinson, S. C., Enfield, N. J., Gaby, A., & Majid, A. (2004). "Reciprocal constructions and situation type." In A. Majid (Ed.), *Field manual.* Vol. 9 (pp. 25–30). Nijmegen: Max Planck Institute for Psycholinguistics.

Evans, N., Gaby, A., & Nordlinger, R. (2007). "Valency mismatches and the coding of reciprocity in Australian languages." *Linguistic Typology, 11*(2), 543–599.

Evans, N., Gaby, A., Levinson, S. C., & Majid, A. (Eds.) (2011a). *Reciprocals and semantic typology.* Amsterdam: John Benjamins.

Evans, N., Levinson, S. C., Gaby, A., & Majid, A. (2011b). "Introduction: Reciprocals and semantic typology." In N. Evans, A. Gaby, S. C. Levinson, & A. Majid (Eds.), *Reciprocals and semantic typology* (pp. 1–28). Amsterdam: John Benjamins.

Everett, D. L. (2004). "Cultural constraints on grammar and cognition in Pirahã: Another look at the design features of human language." *Current Anthropology, 46*(4), 621–646.

———. (2008). *Don't sleep, there are snakes: Life and language in the Amazonian jungle.* New York: Pantheon.

———. (2012). *Language: The cultural tool.* London: Profile Books.

———. (2016). *Dark matter of the mind.* Chicago: University of Chicago Press.

———. (2017). *How language began: The story of humanity's greatest invention.* London: Profile Books.

Ewert, J. -P. (1974). "The neural basis of visually guided behavior." *Scientific American, 230*(3), 34–42.

———. (2004). "Motion perception shapes the visual world of amphibians." In F. R. Prete (Ed.), *Complex worlds from simpler nervous systems* (pp. 117–160). Cambridge, MA: The MIT Press.

Fehr, E., & Fischbacher, U. (2004). "Social norms and human cooperation." *Trends in Cognitive Sciences, 8*(4), 185–190.

Finlay, B. L., Darlington, R. B., & Nicastro, N. (2001). "Developmental structure in brain evolution." *Behavioral and Brain Sciences, 24,* 263–308.

Fisher, C., & Tokura, H. (1996). "Acoustic cues to grammatical structure in infant-directed speech: Cross-linguistic evidence." *Child Development, 67,* 3192–3218.

Flierl, W., & Strauss, H. (Eds.) (1977). *Kâte dictionary.* Series C-41. Canberra: Pacific Linguistics.

Fodor, J. (1983). *The modularity of mind.* Cambridge, MA: The MIT Press.

Foley, W. A. (1986). *The Papuan languages of New Guinea.* Cambridge: Cambridge University Press.

———. (1991). *The Yimas language of New Guinea.* Palo Alto, CA: Stanford University Press.

Frajzyngier, Z., & Curl, T. S. (Eds.) (1999). *Reciprocals: Forms and functions.* Amsterdam: John Benjamins.

Friederici, A. D. (2002). "Towards a neural basis of auditory sentences processing." *Trends in Cognitive Sciences, 6*(2), 78–84.

———. (2005). "Neurophysiological markers of early language acquisition: From syllables to sentences." *Trends in Cognitive Sciences, 9*(10), 481–488.

———. (2009). "Pathways to language: Fiber tracts in the human brain." *Trends in Cognitive Sciences, 13*(4), 175–181.

Friederici, A. D., Fiebach, C. J., Schlesewsky, M., Bornkessel, I. D., & von Cramon, D. Y. (2006). "Processing linguistic complexity and grammaticality in the left frontal cortex." *Cerebral Cortex, 16,* 1709–1717.

Friederici A. D., Mueller, J. L., & Oberecker, R. (2011). "Precursors to natural grammar learning: Preliminary evidence from 4-month-old infants." *PLoS ONE, 6*(3), e17920, 1–7.

Friedrich, M., & Friederici, A. D. (2005). "Phonotactic knowledge and lexical-semantic processing in one-year-olds: Brain responses to words and nonsense words in picture contexts." *Journal of Cognitive Neuroscience, 17,* 1785–1802.

Gärdenfors, P. (2004). "Cooperation and the evolution of symbolic communication." In D. Kimbrough Oller & U. Gabriel (Eds.), *Evolution of communication systems. A comparative approach* (pp. 237–56). Cambridge, MA: The MIT Press.

Gallistel, C. R. (1990). *The organization of learning.* Cambridge, MA: The MIT Press.

———. (1999). "The replacement of general-purpose learning models with adaptively specialized learning modules." In M. S. Gazzangia (Ed.), *The cognitive neurosciences,* second edition (pp. 1179–1191). Cambridge, MA: The MIT Press.

Garson, J. (2012 [1997]). "Connectionism." In E. N. Zalta (Ed.), *The stanford encyclopedia of philosophy,* winter 2012 edition. http://plato.stanford.edu/archives/win2012/entries/connectionism/.

Geary, D. C. (2005). *The origin of mind: Evolution of brain, cognition, and general intelligence.* Washington, DC: American Psychological Association.

Gerhart, J., & Kirschner, M. (2007). "The theory of facilitated variation." *Proceedings of the National Academy of Sciences, 104*(suppl1), 8582–8589.

Gervain, J., & Werker, J. F. (2013). "Prosody cues word order in 7-month-old bilingual infants." *Nature Communications, 4,* 1490.

Ghiglieri, M. P. (1984). *The chimpanzees of Kibale Forest.* New York: Columbia University Press.

Gibson, K. R. (1993). "Tool use, language and social behavior in relationship to information processing capacities." In K. R. Gibson & T. Ingold (Eds.), *Tools, language and cognition in human evolution* (pp. 251–269). Cambridge: Cambridge University Press.

Gibson, K. R., Rumbaugh, D. M., & Beran, M. J. (2001). "Bigger is better: Primate brain size in relationship to cognition." In D. Falk & K. R. Gibson (Eds.), *Evolutionary anatomy of the primate cerebral cortex* (pp. 79–97). Cambridge: Cambridge University Press.

Giurfa, M. (2004). "Conditioning procedure and color discrimination in the honeybee Apis mellifera." *Naturwissenschaften, 91,* 228–231.

Giurfa, M., Eichmann, B., & Menzel, R. (1996). "Symmetry perception in an insect." *Nature, 382,* 458–461.

Giurfa, M., Hammer, M., Stach, S., Stollhoff, N., Müller-Deisig, N., & Mizyrychi, C. (1999). "Pattern learning by honeybees: Conditioning procedures and recognition strategy." *Animal Behaviour, 57,* 315–324.

Giurfa, M., Zhang, S., Jenett, A., Menzel, R., & Srinivasan, M. V. (2001). "The concepts of 'sameness' and 'difference' in an insect." *Nature, 410,* 930–933.

Givón, T. (1985). "Iconicity, isomorphism and non-arbitrary coding in syntax." In J. Haiman (Ed.), *Iconicity in syntax* (pp. 187–219). Amsterdam: John Benjamins.

———. (1991). "Isomorphism in the grammatical code: Cognitive and biological considerations." *Studies in Language, 15*(1), 85–114.

Gleitman, L. (1990). "The structural sources of verb meanings." *Language Acquisition, 1,* 3–55.

Good, J. (2008). "Introduction." In J. Good (Ed.), *Linguistic universals and language change* (pp. 1–19). Oxford: Oxford University Press.

Goodall, J. (1970). "Tool-using in primates and other vertebrates." In D. S. Lehrmann, R. A. Hinde, & E. Shaw (Eds.), *Advances in the study of behavior.* Vol. 3 (pp. 195–249). New York: Academic Press.

———. (1986). *The chimpanzees from Gombe: Patterns of behavior.* Cambridge, MA: Harvard University Press.

Goodman, C. S., & Coughlin, B. S. (Eds.) (2000). "Special feature: The evolution of evo-devo biology." *Proceedings of the National Academy of Sciences, 97*(9), 4424–4456.

Goody, E. N. (1997). "Social intelligence and language: Another rubicon?" In A. Whiten & R. W. Byrne (Eds.), *Machiavellian intelligence II: Extensions and evaluations* (pp. 365–96). Cambridge: Cambridge University Press.

Goren, C. C., Sarty, M., & Wu, P. Y. K. (1975). "Visual following and pattern discrimination of face-like stimuli by newborn infants." *Pediatrics, 56*(4), 544–549.

Gould, S. J. (1977). *Ontogeny and phylogeny.* Cambridge, MA: Harvard University Press.

Gould, S. J., & Lewontin, R. (1979). "The spandrels of San Marco and the Panglossian paradigm: A critique of the adaptationist programme." *Proceedings of the Royal Society B, 205*(1161), 581–598.

Gómez, R., & Maye, J. (2005). "The developmental trajectory of nonadjacent dependency learning." *Infancy, 7,* 183–206.

Greenfield, P. M. (1991). "Language, tools and brain: The ontogeny and phylogeny of hierarchically organized sequential behavior." *Behavioral and Brain Sciences, 14,* 531–595.

Haiman, J. (1983). "Iconic and economic motivation." *Language, 59,* 781–819.

———. (2000). "Iconicity." In G. Booij, J. Mugdan, & C. Lehmann (Eds.), *Morphology: An international handbook.* Vol. 1 (pp. 281–288). Berlin: de Gruyter.

Hale, K. (1976). "The adjoined relative clause in Australian languages." In R. M. W. Dixon (Ed.), *Grammatical categories in Australian languages* (pp. 78–105). Canberra: Australian Institute of Aboriginal Studies.

———. (1992). "Basic word order in two 'free word order' languages." In D. L. Payne (Ed.), *Pragmatics of word order flexibility* (pp. 63–82). Amsterdam: John Benjamins.

Hamilton, W. D. (1964). "The genetic evolution of social behavior." *Journal of Theoretical Biology, 7,* 1–52.

Harvey, M. (2003). "An initial reconstruction of Proto Gunwinyguan phonology." In N. Evans (Ed.), *The non-Pama-Nyungan languages of Northern Australia. Comparative studies of the continent's most linguistically complex region* (pp. 204–268). Canberra: Pacific Linguistics.

Haspelmath, M. (2000). "Why can't we talk to each other? A review article of [Newmeyer, Frederick. 1998. Language form and language function. Cambridge, MA: MIT Press.]" *Lingua, 110*(4), 235–255.

———. (2001). "The European linguistic area: Standard Average European." In M. Haspelmath, E. König, W. Oesterreicher, & W. Raible (Eds.), *Language typology and language universals* (pp. 1492–1510). Berlin: de Gruyter.

———. (2007a). "Pre-established categories don't exist: Consequences for language description and typology." *Linguistic Typology, 11*(1), 119–132.

———. (2007b). "Further remarks on reciprocal constructions." In V. P. Nedjalkov (Ed.), *Reciprocal constructions* (pp. 2087–2115). Amsterdam: John Benjamins.

———. (2008a). "Parametric versus functional explanations of syntactic universals." In T. Biberauer (Ed.), *The limits of syntactic variation* (pp. 75–107). Amsterdam: John Benjamins.

———. (2008b). "Frequency vs. iconicity in explaining grammatical asymmetries." *Cognitive Linguistics, 19*(1), 1–33.

Hauser, M., Chomsky, N., & Fitch, W. T. (2002). "The faculty of language: What is it, who has it, and how did it evolve?" *Science, 298,* 1569–1579.

Hawkes, K. (2012). "Stag hunts or rearing environments? Comment on [M. Tomasello et al. 2012. Two key steps in the evolution of human cooperation: The interdependence hypothesis.]" *Current Anthropology, 53*(6), 687–688.

Hawkins, J. A. (2004). *Efficiency and complexity in grammars.* Oxford: Oxford University Press.

Heath, J. (1980). *Nunggubuyu myths and ethnographic texts.* Canberra: Australian Institute of Aboriginal Studies, AIAS.

———. (1981). *Nunggubuyu dictionary.* Canberra: Australian Institute of Aboriginal Studies, AIAS.

———. (1984). *Functional grammar of Nunggubuyu.* Canberra: Australian Institute of Aboriginal Studies, AIAS.

Heine, B., & Miyashita, H. (2008). "The intersection between reflexives and reciprocals: A grammaticalization perspective." In E. König & V. Gast (Eds.), *Reciprocals and reflexives: Theoretical and typological explorations* (pp. 169–223). Berlin: Mouton de Gruyter.

Heine, B., Claudi, U., & Hünnemeyer, F. (1991). *Grammaticalization: A conceptual framework.* Chicago: University of Chicago Press.

Herculano-Houzel, S. (2016). *The human advantage: A new understanding of how our brain became remarkable.* Cambridge, MA: The MIT Press.

Herrmann, B., Thöni, C., & Gächter, S. (2008). "Antisocial punishment across societies." *Science, 319*(5868), 1362–1367.

Herrnstein, R. J., Loveland, D., & Cable, C. (1976). "Natural concepts in pigeons." *Journal of Experimental Psychology: Animal Behavior Processes, 2,* 285–302.

Hicks, D. (2010). "The material cultural turn: Event and effect." In D. Hicks & M. C. Beaudry (Eds.), *The Oxford handbook of material culture studies* (pp. 25–98). Oxford: Oxford University Press.

Hockett, C. F. (1960). "The origin of speech." *Scientific American, 203*(3), 89–96.

Hockett, C. F., & Altmann, S. A. (1968). "A note on design features." In T. A. Sebeok (Ed.), *Animal communication: Techniques of study and results of research* (pp. 61–72). Bloomington: Indiana University Press.

Höhle, B. (2009). "Bootstrapping mechanisms in first language acquisition." *Linguistics, 47*(2), 359–382.

Hopper, P. J. (1987). "Emergent grammar." *Berkeley Linguistic Society, 13,* 139–157.

Hopper. P. (2001). "Grammatical constructions and their discourse origins: Prototype or family resemblance?" In M. Pütz, S. Niemeier, & R. Dirven (Eds.), *Applied cognitive linguistics I: Theory and language acquisition* (pp. 109–130). Berlin: Mouton de Gruyter.

Hopper, L. M., Spiteri, A., Lambeth, S. P., Schapiro, S. J., Horner, V., & Whiten, A. (2007). "Experimental studies of traditions and underlying transmission processes in chimpanzees." *Animal Behaviour, 73,* 1021–1032.

Hopper, P., & Traugott, E. C. (1993). *Grammaticalization.* Cambridge: Cambridge University Press.

Horner, V., Proctor, D., Bonnie, K. E., Whiten, A., & De Waal, F. B. M. (2010). "Prestige affects cultural learning in chimpanzees." *PLoS ONE, 5*(5), e10625.

Hosokawa, K. (1991). "The Yawuru language of West Kimberley: A meaning-based description." PhD thesis, Canberra, Australian National University.

Huang, C. -T. J. (1984). "On the distribution and reference of empty pronouns." *Linguistic Inquiry, 15,* 531–574.

Hubley, P., & Trevarthen, C. (1979). "Sharing a task in infancy." *New Directions for Child and Adolescent Development, 4,* 57–80.

Humle, T. (2011a). "Ant-dipping: How ants have shed light on culture." In T. Matsuzawa, T. Humle, & Y. Sugiyama (Eds.), *The chimpanzees of Bossou and Nimba* (pp. 97–105). Tokyo: Springer.

———. (2011b). "The chimpanzees from Yealé, Nimba." In T. Matsuzawa, T. Humle, & Y. Sugiyama (Eds.), *The chimpanzees of Bossou and Nimba* (pp. 267–275). Tokyo: Springer.

Humle, T., & Matsuzawa, T. (2004). "Oil palm use by adjacent communities of chimpanzees at Bossou and Nimba Mountains, West Africa." *International Journal of Primatology, 25,* 551–581.

Hurford, J. R. (2007). *The origins of meaning: Language in the light of evolution.* Oxford: Oxford University Press.

———. (2012). *The origins of grammar: Language in the light of evolution.* Oxford: Oxford University Press.

Hyslop, C. (2001). *The Lolovoli dialect of the North-east Ambae language, Vanuatu.* Canberra: Pacific Linguistics.

Ibbotson, P., & Tomasello, M. (2016). "Evidence rebuts Chomsky's theory of language learning." *Scientific American, 315*(5), 70–75.

Ingold, T. (1993). "Tool-use, sociality and intelligence." In K. R. Gibson & T. Ingold (Eds.), *Tools, language and cognition in human evolution* (pp. 429–445). Cambridge: Cambridge University Press.

Jackendoff, R. (2002). *Foundations of language: Brain, meaning, grammar, evolution.* Oxford: Oxford University Press.

———. (2007). *Language, consciousness, culture: Essays on mental structure.* Cambridge, MA: The MIT Press.

Johnson, M. H., Dziurawiec, S., Ellis, H., & Morton, J. (1991). "Newborns' preferential tracking of face-like stimuli and its subsequent decline." *Cognition*, 40 (1–2), 1–19.

Johnson, E. K., & Jusczyk, P. W. (2001). "Word segmentation by 8-month-olds: When speech cues count more than statistics." *Journal of Memory and Language, 44,* 548–567.

Johnson, D. E., & Lappin, S. (1997). "A critique of the minimalist program." *Linguistics and Philosophy, 20,* 273–333.

Johnson, D. E., & Postal, P. (1980). *Arc pair grammar.* Princeton, NJ: Princeton University Press.

Jusczyk, P. W. (2000 [1997]). *The discovery of spoken language.* Cambridge, MA: The MIT Press.

Jusczyk, P. W., Hirsh-Pasek, K., Kemler Nelson, D. G., Kennedy, L. J., Woodward, A., & Piwoz, J. (1992). "Perception of acoustic correlates of major phrasal units by young infants." *Cognitive Psychology, 24,* 252–293.

Kaplan, R., & Bresnan, J. (1982). "Lexical-functional grammar: A formal system for grammatical representation." In J. Bresnan (Ed.), *The mental representation of grammatical relations* (pp. 173–281). Cambridge, MA: The MIT Press.

Kay, M. (1983). *Unification grammar.* Technical Report, Palo Alto, CA: Xerox Palo Alto Research Center.

Keller, R. (1994). *On language change: The invisible hand in language.* London: Routledge.

Kemmer, S. (1993). *The middle voice.* Amsterdam: John Benjamins.

Kemmer, S., & Barlow, M. (2000). "Introduction: A usage-based conception of language." In M. Barlow & S. Kemmer (Eds.), *Usage-based models of language* (pp. vii–xxvii). Stanford, CA: CSLI Publications.

Keyßer, C. (1969 [1925]). "Wörterbuch der Kâte-Sprache gesprochen in Neuguinea. (Dictionary of the Kâte language as spoken in New Guinea.)" In C. Meinhof (Ed.), *Beihefte zur Zeitschrift für Eingeborenen-Sprachen.* Siebtes Heft [Issue 7]. Berlin: Verlag von Dietrich Reimer (Ernst Vohsen) (1925); Kraus Reprint (1969).

Kite, S., & Wurm, S. (2004). *The Duuŋidjawu language of Southeast Queensland: Grammar, texts and vocabulary.* Canberra: Pacific Linguistics.

Klatt, D. H. (1976). "Linguistic uses of segment duration in English: Acoustic and perceptual evidence." *Journal of Phonetics, 3,* 129–140.

Knjazev, J. P. (2007). "Lexical reciprocals as a means of expressing reciprocal situations." In V. P. Nedjalkov (Ed.), *Reciprocal constructions* (pp. 115–146). Amsterdam: John Benjamins.

Knudsen, E. I. (1999). "Mechanisms of experience-dependent plasticity in the auditory localization pathway of the barn owl." *Journal of Comparative Physiology. A, Sensory, Neural and Behavioural Physiology, 185,* 305–321.

Köhler, W. (1921 [1963]). *Intelligenzprüfungen an Menschenaffen* [Studies on intelligence in great apes = The mentality of apes]. Berlin: Springer.

———. (1957 [1925]). *The mentality of apes.* Harmondsworth: Pelican Books.

———. (1925 [1976]). *The mentality of apes.* London: Routledge & Kegan Paul (1925); New York: Liveright (1976).

———. (1966 [1938]). *The place of value in a world of facts.* New York: Liveright.

Kohli, M., & Künemund, H. (2003). "Intergenerational transfers in the family. What motivates giving?" In V. L. Bengtson & A. Lowenstein (Eds.), *Global aging and challenges to families* (pp. 123–142). New York: Aldine de Gruyter.

König, E. (2005). "Reciprocity in language: Cultural concepts and patterns of encoding." *Uhlenbeck Lecture, 23,* 1–26.

———. (2011). "Reciprocals and semantic typology: Some concluding remarks." In N. Evans, A. Gaby, S. C. Levinson, & A. Majid (Eds.), *Reciprocals and semantic typology* (pp. 329–339). Amsterdam: John Benjamins.

König, E., & Gast, V. (Eds.) (2008a). *Reciprocals and reflexives: Theoretical and typological explorations.* Berlin: Mouton de Gruyter.

———. (2008b). "Reciprocity and reflexivity: Description, typology and theory." In E. König & V. Gast (Eds.), *Reciprocals and reflexives: Theoretical and typological explorations* (pp. 1–31). Berlin: Mouton de Gruyter.

König, E., & Kokutani, S. (2006). "Towards a typology of reciprocal constructions: Focus on German and Japanese." *Linguistics, 44(2),* 271–302.

König, E., & Vezzosi, L. (2004). "The role of predicate meaning in the development of reflexivity." In W. Bisang, N. Himmelmann, & B. Wiener (Eds.), *What makes grammaticalization? A look from its fringes and its components* (pp. 213–244). Berlin: Mouton de Gruyter.

Kolodny, O., & Edelman, S. (2018). "The evolution of the capacity for language: The ecological context and adaptive value of a process of cognitive hijacking." *Philosophical Transactions of the Royal Society B, 373,* 20170052.

Koops, K. (2011). "Chimpanzees in the Seringbara region of the Nimba Mountains." In T. Matsuzawa, T. Humle, & Y. Sugiyama (Eds.), *The chimpanzees of Bossou and Nimba* (pp. 277–287). Tokyo: Springer.

Koops, K., McGrew, W. C., & Matsuzawa, T. (2010). "Do chimpanzees (*Pan troglodytes*) use cleavers and anvils to facture *Treculia africana* fruits? Preliminary data on a new form of percussive technology." *Primates, 51,* 175–178.

Kruspe, N. (2011). "Reciprocal constructions in Mah Meri." In N. Evans, A. Gaby, S. C. Levinson, & A. Majid (Eds.), *Reciprocals and semantic typology* (pp. 149–162). Amsterdam: John Benjamins.

Kube, S., Maréchal, M. A., & Puppe, C. (2012). "The currency of reciprocity: Gift exchange in the workplace." *American Economic Review, 102*(4), 1644–1662.

Kuhl, P. K. (2004). "Early language acquisition: Cracking the speech code." *Nature Reviews Neuroscience, 5,* 831–843.

———. (2010). "Brain mechanisms in early language acquisition." *Neuron, 67,* 713–727.

Kuhle, A. (2007). "Reziprozität aus sprachtypologischer Perspektive und ihre Beziehungen zu nicht-linguistischen Konzepten" [Reciprocity from a language-typological perspective and its relationship to non-linguistic concepts]. M.A. thesis, Freie Universität Berlin. Unpublished manuscript.

———. (2018a). "(Dis)Continuity: The cultural intelligence hypothesis reconsidered." *Language Sciences, 67,* 59–71.

———. (2018b). "Clues to the puzzle: The significance of material cultures in nonhuman primates for the study of language." *Language Sciences, 67,* 72–88.

Landau, B., & Gleitman, L. (1985). *Language and experience: Evidence from the blind child.* Cambridge, MA: Harvard University Press.

Lehmann, C. (1995 [1982]). *Thoughts on grammaticalization.* Munich: Lincom Europa.

Levinson, S. C. (2003). *Space in language and cognition: Explorations in cognitive diversity.* New York: Cambridge University Press.

Levinson, S. C., & Wilkins, D. P. (Eds.) (2006). *Grammars of space: Explorations in cognitive diversity.* New York: Cambridge University Press.

Lewis, D. (1969). *Convention.* Cambridge, MA: Harvard University Press.

Lichtenberk, F. (1983a). *A grammar of Manam.* Honolulu: University of Hawai'i Press.

———. (1983b). "Relational classifiers." *Lingua, 60,* 147–176.

———. (1985). "Multiple uses of reciprocal constructions." *Australian Journal of Linguistics, 5,* 19–41.

———. (1999). "Reciprocals without reflexives." In Z. Frajzyngier & T. S. Curl (Eds.), *Reciprocals: Forms and functions* (pp. 31–62). Amsterdam: John Benjamins.

———. (2009). "Attributive possessive constructions in Oceanic." In W. B. McGregor (Ed.), *The expression of possession* (pp. 249–292). Berlin: Mouton de Gruyter.

Lichtenberk, F., Vaid, J., & Chen, H. -C. (2011). "On the interpretation of alienable vs. inalienable possession: A psycholinguistic investigation." *Cognitive Linguistics, 22*(4), 659–689.

Linell, P. (2005). *The written language bias in linguistics: Its nature, origins and transformations.* London: Routledge.

Logothetis, N. K., & Sheinberg, D. L. (1996). "Visual object recognition." *Annual Review of Neuroscience, 19,* 577–621.

Lorenz, K. (1981). *The foundations of ethology.* New York: Springer.

Lorenz, K., & Tinbergen, N. (1938). "Taxis und Instinkthandlung in der Eieinrollbewegung der Graugans." *Zeitschrift für Tierpsychologie, 2,* 1–29.

Love, N. (2004). "Cognition and the language myth." *Language Sciences, 26*(6), 525–544.

Lucy, J. A. (1992). *Language diversity and thought: A reformulation of the linguistic relativity hypothesis.* New York: Cambridge University Press.

———. (1996a). *Grammatical categories and cognition: A case study of the linguistic relativity hypothesis.* New York: Cambridge University Press.

———. (1996b). "The scope of linguistic relativity: An analysis and review of empirical research." In J. J. Gumperz & S. C. Levinson (Eds.), *Rethinking linguistic relativity* (pp. 37–69). Cambridge: Cambridge University Press.

Luncz, L. V., Mundry, R., & Boesch, C. (2012). "Evidence for cultural differences between neighboring chimpanzee communities." *Current Biology, 22,* 1–5.

Lust, B. C., Wali, K., Gair, J. W., & Subbarao, K. V. (Eds.) (2000). *Lexical anaphors and pronouns in selected South Asian languages: A principled typology.* Berlin: Mouton de Gruyter.

Lynch, J. (1998). *Pacific Languages: An introduction.* Honolulu: University of Hawai'i Press.

Lyons, J. (1977). *Semantics.* Vol. 2. Cambridge: Cambridge University Press.

Mahmoudzadeh, M., Dehaene-Lambertz, G., Fournier, M., Kongolo, G., Goudjil, S., Dubois, J., Grebe, R., & Wallois, F. (2013). "Syllabic discrimination in premature human infants prior to complete formation of cortical layers." *Proceedings of the National Academy of Sciences, 110*(12), 4846–4851.

Majid, A., Evans, N., Gaby, A., & Levinson, S. C. (2011a). "The grammar of exchange: A comparative study of reciprocal constructions across languages." *Frontiers in Psychology, 2*(34), 1–15.

———. (2011b). "The semantics of reciprocal constructions across languages: An extensional approach." In N. Evans, A. Gaby, S. C. Levinson, & A. Majid (Eds.), *Reciprocals and semantic typology* (pp. 29–60). Amsterdam: John Benjamins.

Malle, B. F., Moses, L. J., & Baldwin, D. A. (2001). "Introduction: The significance of intentionality." In B. F. Malle, L. J. Moses, & D. A. Baldwin (Eds.), *Intentions and intentionality: Foundations of social cognition* (pp. 1–24). Cambridge, MA: The MIT Press.

Marino, L. (2006). "Absolute brain size: Did we throw out the baby with the bathwater?" *Proceedings of the National Academy of Sciences, USA, 103,* 13563–13564.

Maslova, E. (2007a). "Reciprocal and polyadic (remarkable reciprocals in Bantu)." In V. P. Nedjalkov (Ed.), *Reciprocal constructions* (pp. 335–352). Amsterdam: John Benjamins.

———. (2007b). "Reciprocals in Yukaghir languages." In V. P. Nedjalkov (Ed.), *Reciprocal constructions* (pp. 1835–1863). Amsterdam: John Benjamins.

———. (2008). "Reflexive encoding of reciprocity: Cross-linguistic and language-internal variation." In E. König & V. Gast (Eds.), *Reciprocals and reflexives: Theoretical and typological explorations* (pp. 225–257). Berlin: Mouton de Gruyter.

Maslova, E., & Nedjalkov, V. P. (2005). "Reciprocal constructions." In M. Haspelmath, M. Dryer, D. Gil, & B. Comrie (Eds.), *The world atlas of language structures* (pp. 430–433). Oxford: Oxford University Press.

———. (2013). "Reciprocal constructions." In M. S. Dryer & M. Haspelmath (Eds.), *The world atlas of language structures online*. Leipzig: Max Planck Institute for Evolutionary Anthropology. (Available online at http://wals.info/chapter/106, Accessed on 2017-09-06.)

Matsuzawa, T. (2011). "Field experiments of tool-use." In T. Matsuzawa, T. Humle, & Y. Sugiyama (Eds.), *The chimpanzees of Bossou and Nimba* (pp. 157–164). Tokyo: Springer.

Matsuzawa, T., Humle, T., & Sugiyama, Y. (Eds.) (2011). *The chimpanzees of Bossou and Nimba.* Tokyo: Springer.

Matsuzawa, T., & Yamakoshi, G. (1996). "Comparison of chimpanzee material culture between Bossou and Nimba, West Africa." In A. Russon, K. Bord, & S. Parho (Eds.), *Reaching into thought: The mind of the great apes* (pp. 211–232). Cambridge: Cambridge University Press.

Mattys, S. L., Jusczyk, P. W., Luce, P. A., & Morgan, J. L. (1999). "Phonotactic and prosodic effects on word segmentation in infants." *Cognitive Psychology, 38,* 465–494.

Mauss, M. (1990 [1924]). *The gift: Forms and functions of exchange in archaic societies.* New York: W. W. Norton & Company.

Maye, J., Werker, J. F., & Gerken, L. (2002). "Infant sensitivity to distributional information can affect phonetic discrimination." *Cognition, 82,* B101–B111.

Mayer, R. E. (1995). "The search for insight: Grappling with gestalt psychology's unanswered questions." In R. J. Sternberg & J. E. Davidson (Eds.), *The nature of insight* (pp. 3–32). Cambridge, MA: The MIT Press.

McElhanon, K. (1975). "North-eastern Trans-New Guinea Phylum languages." *Pacific Linguistics, C38,* 527–568.

McGregor, W. B. (1999). "Reflexive and reciprocal constructions in Nyulnyulan languages." In Z. Frajzyngier & T. S. Curl (Eds.), *Reciprocals: Forms and functions* (pp. 85–122). Amsterdam: John Benjamins.

———. (2002). *Verb classification in Australian languages.* Berlin: Mouton de Gruyter.

McGregor, W. (2009). "Introduction." In W. B. McGregor (Ed.), *The expression of possession* (pp. 1–12). Berlin: Mouton de Gruyter.

McGrew, W. C. (1974). "Tool use by wild chimpanzees in feeding upon driver ants." *Journal of Human Evolution, 3,* 501–508.

———. (1998). "Culture in nonhuman primates?" *Annual Review of Anthropology, 27,* 301–328.

McGrew, W. C., Ham, R. M., White, L. J. T., Tutin, C. E. G., & Fernandez, M. (1997). "Why don't chimpanzees in Gabon crack nuts?" *International Journal of Primatology, 18*(3), 353–374.

McGuckin, C. (2002). "Gapapaiwa." In J. Lynch, M. Ross, & T. Crowley (Eds.), *The Oceanic languages* (pp. 297–321). Richmond, Surrey: Curzon.

McQueen, J. M. (1998). "Segmentation of continuous speech using phonotactics." *Journal of Memory and Language, 39,* 21–46.

Melis, A. P., & Tomasello, M. (2013). "Chimpanzees' (Pan troglodytes) strategic helping in a collaborative task." *Biology Letters, 9,* 20130009. doi: 10.1098/rsbl.2013.0009.

Menzel, R. (2013). "Learning, memory, and cognition: Animal perspectives." In G. C. Galizia & P. M. Liedo (Eds.), *Neurosciences – From molecule to behavior: A university textbook* (pp. 629–653). Heidelberg: Springer Spektrum.

———. (2017). "Search strategies for intentionality in the honeybee brain." In J. H. Byrne (Ed.), *The Oxford handbook of invertebrate neurobiology.* Oxford: Oxford University Press. Online publication.

Menzel, R., Brandt, R., Gumbert, A., Komischke, B., & Kunze, J. (2000). "Two spatial memories for honeybee navigation." *Proceedings of the Royal Society of London B, 267,* 961–968.

Menzel, R., & Giurfa, M. (2006). "Dimensions of cognition in an insect, the honeybee." *Behavioral and Cognitive Neuroscience Reviews, 5,* 24–40.

Menzel, R., Greggers, U., Smith, A., Berger, S., Brandt, R., Brunke, S., Bundrock, G., Hülse, S., Plümpe, T., Schaupp, F., Schüttler, E., Stach, S., Stindt, J., Stollhoff, N., & Watzl, S. (2005). "Honybees navigate according to a map-like spatial memory." *Proceedings of the National Academy of Sciences, 102*(8), 3040–3045.

Merlan, F., & Rumsey, A. (1989). *Ku Waru – English word list.* Unpublished manuscript.

———. (1991). *Ku Waru: Language and segmentary politics in the western Nebilyer Valley, Papua New Guinea.* Cambridge: Cambridge University Press.

Mithun, M. (1984). "How to avoid subordination." *Berkeley Linguistic Society, 10,* 493–509.

Moll, H., & Tomasello, M. (2007). "Cooperation and human cognition: The Vygotskian intelligence hypothesis." *Philosophical Transactions of the Royal Society B, 362,* 639–648.

Molm, L. D. (2010). "The structure of reciprocity." *Social Psychology Quarterly, 73*(2), 119–131.

Morgan, B. J., & Abwe, E. E. (2006). "Chimpanzees use stone hammers in Cameroon." *Current Biology, 16,* 632–633.

Mueller, J. L., Friederici, A. D., & Männel, C. (2012). "Auditory perception at the root of language learning." *Proceedings of the National Academy of Sciences of the United States of America, 109*(39), 15953–15958.

Müller, G. B., & Newman, S. A. (Eds.) (2005). "Special issue: Evolutionary innovation and morphological novelty." *Journal of Experimental Zoology Part B: Molecular and Developmental Evolution, 304B,* 485–631.

Muysken, P. (1981). "Quechua causatives and logical form: A case study of markedness." In A. Belletti, L. Brandi, & L. Rizzi (Eds.), *Theory of markedness in generative grammar* (pp. 445–473). Pisa: Scuola Normale Superiore di Pisa.

Nedjalkov, V. P. (Ed.) (2007a). *Reciprocal constructions.* Amsterdam: John Benjamins.

———. (2007b). "Overview of the research: Definitions of terms, framework, and related issues." In V. P. Nedjalkov (Ed.), *Reciprocal constructions* (pp. 3–114). Amsterdam: John Benjamins.

Nedjalkov, V. P., & Geniušienė, E. Š. (2007). "Questionnaire on reciprocals." In V. P. Nedjalkov (Ed.), *Reciprocal constructions* (pp. 379–434). Amsterdam: John Benjamins.

Nekes, H., & Worms, E. A. (1953). *Australian languages.* Fribourg: Anthropos Institut.

Nishida, T., & Uehara, S. (1983). "Natural diet of chimpanzees (*Pan troglodytes schweinfurthii*): Long-term record from the Mahale Mountains, Tanzania." *African Study Monographs, 3,* 109–130.

Nowak, M. A., & Sigmund, K. (2005). "Evolution of indirect reciprocity." *Nature, 437,* 1291–1298.

Partanen, E., Kujala, T., Näätänen, R., Liitola, A., Sambeth, A., & Huotilainen, M. (2013). "Learning-induced neural plasticity of speech processing before birth." *Proceedings of the National Academy of Sciences, 110*(37), 15145–15150.

Patz, E. (1991). "Djabugay." In R. M. W. Dixon & B. J. Blake (Eds.), *The handbook of Australian languages.* Vol. 4 (pp. 245–347). Oxford: Oxford University Press.

Peirce, C. S. (1932). *Collected papers of C. S. Peirce.* Vol. 2. Cambridge, MA: Harvard University Press.

Petitto, L. -A. (1999). "The acquisition of natural signed languages: Lessons in the nature of human language and its biological foundations." In C. Chamberlain, J. Morford, & R. Mayberry (Eds.), *Language acquisition by eye* (pp. 41–50). Mahwah, NJ: Lawrence Erlbaum.

Piaget, J. (1973). *Main trends in inter-disciplinary research.* New York: Harper & Row.

Piattelli-Palmarini, M. (2009). "What is language, that it may have evolved, and what is evolution, that it may apply to language." In R. K. Larson, V. Deprez, & H. Yamakido (Eds.), *The evolution of human language: Biolinguistic perspectives* (pp. 148–162). Cambridge: Cambridge University Press.

Piattelli-Palmarini, M., & Uriagereka, J. (2004). "The immune syntax: The evolution of the language virus." In L. Jenkins (Ed.), *Variations and universals in biolinguistics* (pp. 341–377). Amsterdam: Elsevier North-Holland.

———. (2008). "Still a bridge too far? Biolinguistic questions for grounding language on brains." *Physics of Life Reviews, 5,* 207–224.

Pilhofer, G. (1969 [1933]). "Grammatik der Kâte-Sprache in Neuguinea. [Grammer of the Kâte language in New Guinea]." In C. Meinhof (Ed.), *Beihefte zur Zeitschrift für Eingeborenen-Sprachen.* Vierzehntes Heft [Issue 14]. Berlin: Verlag von Dietrich Reimer (Ernst Vohsen) (1933); Kraus Reprint (1969).

Pinker, S. (1984). *Language learnability and language development.* Cambridge, MA: The MIT Press.

———. (1987). "The bootstrapping problem in language acquisition." In B. MacWhinney (Ed.), *Mechanisms of language acquisition* (pp. 399–442). Hillsdale, NJ: Erlbaum.

———. (1994). *The language instinct.* London: Penguin Books.

Pinker, S., & Bloom, P. (1990). "Natural language and natural selection." *Behavioral and Brain Sciences, 13,* 707–784.

Pinker, S., & Jackendoff, R. (2005). "The faculty of language: What's special about it?" *Cognition, 95,* 201–236.

Pollard, C., & Sag, I. (1994). *Head driven phrase structure grammar.* Chicago: University of Chicago Press.

Rapold, C. J. (2011). "Semantics of Khoekhoe reciprocal constructions." In N. Evans, A. Gaby, S. C. Levinson, & A. Majid (Eds.), *Reciprocals and semantic typology* (pp. 61–74). Amsterdam: John Benjamins.

Read, D. W. (2008). "Working memory: A cognitive limit to non-human primate recursive thinking prior to hominid evolution." *Evolutionary Psychology, 6*(4), 676–714.

Reader, S. M., Hager, Y., & Laland, K. N. (2011). "The evolution of primate general and cultural intelligence." *Philosophical Transactions of the Royal Society B, 366,* 1017–1027.

Ross, M. (2005). "Pronouns as a preliminary diagnostic for grouping Papuan languages." In A. Pawley, R. Attenborough, R. Hide, & J. Golson (Eds.), *Papuan pasts: Cultural, linguistic and biological histories of Papuan-speaking peoples* (pp. 15–66). Canberra: Pacific Linguistics.

Rumsey, A. (1990). "Wording, meaning, and linguistic ideology." *American Anthropologist, 92*(2), 346–361.

———. (2010). "Lingual and cultural wholes and fields." In T. Otto & N. Bubandt (Eds.), *Experiments in holism* (pp. 127–149). Oxford: Wiley-Blackwell.

———. (2017). "Dependency and relative determination in language acquisition: The case of Ku Waru." In N. J. Enfield (Ed.), *Dependencies in language: On the causal ontology of linguistic systems* (pp. 97–116). Berlin: Language Science Press.

Russon, A. E., & Begun, D. R. (Eds.) (2004). *The evolution of thought.* Cambridge: Cambridge University Press.

Saffran, J. R., Aslin, R. N., & Newport, E. L. (1996a). "Statistical learning by 8-month-old-infants." *Science, 274,* 1926–1928.

Saffran, J. R., Newport, E. L., & Aslin, R. N. (1996b). "Word segmentation: The role of distributional cues." *Journal of Memory and Language, 35,* 606–621.

Samet, J., & Zaitchik, D. (2017 [2012]). "Innateness and contemporary theories of cognition." In E. N. Zalta (Ed.), *The Stanford encyclopedia of philosophy*, fall 2017 edition. https://plato.stanford.edu/archives/fall2017/entries/innateness-cognition/.

Savage-Rumbaugh, E. S. (1986). *Ape language: From conditioned response to symbol.* New York: Columbia University Press.

Savage-Rumbaugh, E. S., Murphy, J., Sevcik, R. A., Rumbaugh, D. M., Brakke, K. E., & Williams, S. (1993). "Language comprehension in ape and child." *Monographs of the Society for Research in Child Development, 233*(58) (3–4), 1–242.

Savage-Rumbaugh, E. S., Rumbaugh, D. M., & Fields, W. M. (2006). "Language as a window on rationality." In S. Hurley & M. Nudds (Eds.), *Rational animals?* (pp. 513–552). Oxford: Oxford University Press.

Savage-Rumbaugh, E. S., Shanker, S. G., & Taylor, T. J. (1998). *Apes, language, and the human mind.* Oxford: Oxford University Press.

Searle, J. R. (1983). *Intentionality: An essay in the philosophy of mind.* Cambridge: Cambridge University Press.

Seidl, A. (2007). "Infants' use and weighting of prosodic cues in clause segmentation." *Journal of Memory and Language, 57,* 24–48.

Seifert, C. M., Meyer, D. E., Davidson, N., Patalano, A. L., & Yaniv, I. (1995). "Demystification of cognitive insight: Opportunistic assimilation and the prepared-mind perspective." In R. J. Sternberg & J. E. Davidson (Eds.), *The nature of insight* (pp. 65–124). Cambridge, MA: The MIT Press.

Senft, G. (2011). "To have and have not: Kilivila reciprocals." In N. Evans, A. Gaby, S. C. Levinson, & A. Majid (Eds.), *Reciprocals and semantic typology* (pp. 225–232). Amsterdam: John Benjamins.

Sharp, P. E. (1999). "Complimentary roles for hippocampal versus subicular/entorhinal place cells in coding place, context, and events." *Hippocampus, 9,* 432–443.

Siloni, T. (2008). "The syntax of reciprocal verbs: An overview." In E. König & V. Gast (Eds.), *Reciprocals and reflexives: Theoretical and typological explorations* (pp. 451–498). Berlin: Mouton de Gruyter.

Simons, G. F., & Fennig, C. D. (Eds.) (2017). *Ethnologue: Languages of the world, twentieth edition.* Dallas, TX: SIL International. Online version: http://www.ethnologue.com [7/13/2017].

Singer, R. (2011). "Strategies for encoding reciprocity in Mawng." In N. Evans, A. Gaby, S. C. Levinson, & A. Majid (Eds.), *Reciprocals and semantic typology* (pp. 233–250). Amsterdam: John Benjamins.

Soderstrom, M., Seidl, A., Kemler Nelson, D. G., & Jusczyk, P. W. (2003). "The prosodic bootstrapping of phrases: Evidence from prelinguistic infants." *Journal of Memory and Language, 49,* 249–267.

Soderstrom, M., Kemler Nelson, D. G., & Jusczyk, P. W. (2005). "Six-month-olds recognize clauses embedded in different passages of fluent speech." *Infant Behavior and Development, 28,* 87–94.

Steffensen, S. V. (2011). "Beyond mind: An extended ecology of languaging." In S. J. Cowley (Ed,) *Distributed language* (pp. 185–210). Amsterdam: John Benjamins.

———. (2017 [2013]). "Human interactivity: Problem-finding, problem-solving, and verbal patterns in the wild." In S. J. Cowley & F. Vallée-Tourangeau (Eds.), *Cognition beyond the brain: Computation, interactivity and human artifice,* second edition (pp. 85–113). London: Springer.

Steffensen, S. V., & Cowley, S. J. (2010). "Signifying bodies, health and non-locality: The aftermath." In S. J. Cowley, J. M. Major, A. Dinis, & S. V. Steffensen (Eds.), *Signifying bodies: Biosemiosis, interaction and health* (pp. 331–355). Braga: Portuguese Catholic University Press.

Stegbauer, C. (2010 [2002]). *Reziprozität: Einführung in soziale Formen der Gegenseitigkeit* [Reciprocity: Introduction to social forms of exchange]. Second edition. Wiesbaden: Westdeutscher Verlag.

Stokes, B. (1982). "A description of Nyigina: A language of the West Kimberley, Western Australia." PhD thesis, Canberra, Australian National University.

Stokes, B., & McGregor, W. (2003). "Classification and subclassification of the Nyulnyulan languages." In N. Evans (Ed.), *The non-Pama-Nyungan languages of Northern Australia. Comparative studies of the continent's most linguistically complex region* (pp. 29–74). Canberra: Pacific Linguistics.

Stromswold, K. (1995). "The cognitive and neural bases of language acquisition." In Michael S. Gazzaniga (Ed.), *The cognitive neurosciences* (pp. 855–870). Cambridge, MA: The MIT Press.

Suboski, M. D. (1992). "Releaser-induced recognition learning by amphibians and reptiles." *Learning and Behavior, 20*(1), 63–82.

Suddendorf, T., & Corballis, M. C. (1997). "Mental time travel and the evolution of the human mind." *Genetic, Social, and General Psychology Monographs, 123,* 133–167.

Sugiyama, Y. (1985). "The brush-stick of chimpanzees found in south-west Cameroon and their cultural characteristics." *Primates, 26*(4), 361–374.

Sugiyama, Y., & Koman, J. (1979). "Tool-using and making behavior in wild chimpanzees at Bossou, Guinea." *Primates, 20,* 513–524.

Tammero, L. F., & Dickinson, M. H. (2002). "Collision-avoidance and landing responses are mediated by separate pathways in the fruitfly, *Drosophila melanogaster.*" *Journal of Experimental Biology, 205,* 2785–2798.

Taraldsen, T. (1978). *On the NIC, vacuous application, and the that-trace filter.* Indiana Bloomington: University Linguistics Club.

Tarr, M. J., & Cheng, Yi D. (2003). "Learning to see faces and objects." *Trends in Cognitive Sciences, 7,* 23–30.

Tennie, C., Call, J., & Tomasello, M. (2009). "Ratcheting up the ratchet: On the evolution of cumulative culture." *Philosophical Transactions of the Royal Society B: Biological Sciences, 364,* 2405–2415.

Thibault, P. J. (2011). "First-order languaging dynamics and second-order language: The distributed language view." *Ecological Psychology, 23*(3), 210–245.

Thompson, D. W. (1992 [1917]). *On growth and form.* Cambridge: Cambridge University Press.

Thorpe, W. H. (1963). *Learning and instinct in animals.* London: Methuen and Co.

Tinbergen, N. (1951). *The study of instinct.* New York: Oxford University Press.

Tolman, E. C. (1948). "Cognitive maps in rats and men." *Psychological Review, 55,* 189–208.

Tomasello, M. (1990). "Cultural transmission in the tool use and communicatory signaling of chimpanzees?" In S. T. Parker & K. R. Gibson (Eds.), *Language and intelligence in monkeys and apes: Comparative developmental perspectives* (pp. 274–311). New York: Cambridge University Press.

———. (1995). "Language is not an instinct." *Cognitive Development, 10,* 131–156.

———. (1998). "Introduction: A cognitive-functional perspective on language structure." In M. Tomasello (Ed.), *The new psychology of language: Cognitive and functional approaches to language structure.* Vol. 1 (pp. vii–xxiii). Mahwah, NJ: Erlbaum.

———. (1999). *The cultural origins of human cognition.* Cambridge, MA: Harvard University Press.

———. (2003). *Constructing a language: A usage-based theory of language acquisition.* Cambridge, MA: Harvard University Press.

———. (2008). *Origins of human communication.* Cambridge, MA: The MIT Press.

———. (2009). *Why we cooperate.* Based on the 2008 Tanner Lectures on Human Values at Stanford. Cambridge, MA: The MIT Press.

Tomasello, M., Davis-DaSilva, M., Camak, L., & Bard, K. (1987). "Observational learning of tool-use by young chimpanzees." *Human Evolution, 2,* 175–183.

Tomasello, M., Carpenter, M., Call, J., Behne, T., & Moll, H. (2005). "Understanding and sharing intentions: The origins of cultural cognition." *Behavioral and Brain Sciences, 28,* 675–691.

Traugott, E. C. (1990). "From less to more situated in language: The unidirectionality of semantic change." In S. Adamson, V. Law, & N. Vincent (Eds.), *Papers from the fifth international conference on English historical linguistics* (pp. 496–517). Amsterdam: John Benjamins.

Traugott, E. C., & Dasher, R. B. (2002). *Regularity in semantic change.* Cambridge: Cambridge University Press.

Trivers, R. (1971). "The evolution of reciprocal altruism." *Quarterly Review of Biology, 46*, 35–57.

Tsunoda, T. (2007). "Reciprocal constructions in Warrungu." In V. P. Nedjalkov (Ed.), *Reciprocal constructions* (pp. 1403–1434). Amsterdam: John Benjamins.

Tulving, E., & Markowitsch, H. J. (1998). "Episodic and declarative memory: Role of the hippocampus." *Hippocampus, 8,* 198–204.

Vale, G. L., Davis, S. J., Lambeth, S. P., Schapiro, S. J., & Whiten, A. (2017). "Acquisition of a socially learned tool use sequence in chimpanzees: Implications for cumulative culture." *Evolution and Human Behavior, 38,* 635–644.

Van Schaik, C. P., Ancrenaz, M., Borgen, G., Galdikas, B., Knott, C. D., Singleton, I., Suzuki, A., Utami, S. S., & Merrill, M. (2003). "Orangutan cultures and the evolution of material culture." *Science, 299*(102), 102–105.

Van Schaik, C. P., & Knott, C. D. (2001). "Geographic variation in tool use on *Neesia* fruits in orangutans." *American Journal of Physical Anthropology, 114,* 331–342.

Van Staden, M. (2009). "Possessive clauses in East Nusantara, the case of Tidore." In W. B. McGregor (Ed.), *The expression of possession* (pp. 293–341). Berlin: Mouton de Gruyter.

Van Valin, R., & LaPolla, R. J. (1997). *Syntax: Structure, meaning and function.* Cambridge: Cambridge University Press.

Von Mengden, F., & Coussé, E. (2014). "The role of change in usage-based conceptions of language." In E. Coussé & F. von Mengden (Eds.), *Usage-based approaches to language change* (pp. 1–19). Amsterdam: John Benjamins.

Vygotsky, L. S. (1978). *Mind in society: The development of higher psychological processes.* Cambridge, MA: Harvard University Press.

Warneken, F., Chen, F., & Tomasello, M. (2006). "Cooperative activities in young children and chimpanzees." *Child Development, 3,* 640–663.

Warneken, F., Hare, B., Melis, A., Hanus, D., & Tomasello, M. (2007). "Spontaneous altruism by chimpanzees and young children." *PLoS Biology, 5*(7), 1414–1420.

Warneken, F., & Tomasello, M. (2006). "Altruistic helping in human infants and young chimpanzees." *Science, 311*(5765), 1301–1303.

———. (2007). "Helping and cooperation at 14 months of age." *Infancy, 11,* 271–294.

Watanabe, S., Sajamoto, J., & Wakita, M. (1995). "Pigeons' discrimination of paintings by Monet and Picasso." *Journal of Experimental Analysis of Behavior, 63,* 165–174.

Wellmann, C., Holzgrefe1, J., Truckenbrodt, H., Wartenburger, I., & Höhle, B. (2012). "How each prosodic boundary cue matters: Evidence from German infants." *Frontiers in Psychology, 3*(Article 590), 1–13.

Werker, J. F., & Tees, R. C. (1984). "Cross-language speech perception: Evidence for perceptual reorganization during the first year of life." *Infant Behavior and Development, 7,* 49–63.

———. (2005). "Speech perception as a window for understanding plasticity and commitment in language systems of the brain." *Developmental Psychobiology, 46,* 233–251.

Whiten, A. (2005). "The second inheritance system of chimpanzees and humans." *Nature, 437,* 52–55.

———. (2017). "A second inheritance system: The extension of biology through culture." *Interface Focus, 7,* 20160142.

Whiten, A., Goodall, J., McGrew, W. C., Nishida, T., Reynolds, V., Sugiyama, Y., Tutin, C. E. G., Wrangham, R. W., & Boesch, C. (1999). "Cultures in chimpanzees." *Nature, 399,* 682–685.

———. (2001). "Charting cultural variation in chimpanzees." *Behaviour, 138,* 1481–1516.

Whiten, A., Horner, V., & De Waal, F. B. M. (2005). "Conformity to cultural norms of tool use in chimpanzees." *Nature, 437,* 737–740.

Whiten, A., Horner, V., Litchfield, C.A., & Marshall-Pescini, S. (2004). "How do apes ape?" *Learning & Behavior, 32*(1), 36–52.

Whiten, A., McGuigan, N., Marshall-Pescini, S., & Hopper, L.M. (2009). "Emulation, imitation, over-imitation and the scope of culture for child and chimpanzee." *Philosophical Transactions of the Royal Society B, 364,* 2417–2428.

Whiten, A., Spiteri, A., Honer, V., Bonnie, K. E., Lambeth, S. P., Schapiro S. J., & De Waal, F. B. M. (2007). "Transmission of multiple traditions within and between chimpanzee groups." *Current Biology, 17,* 1038–1043.

Whorf, B. L. (1956). *Language, thought, and reality.* Cambridge, MA: The MIT Press.

Wierzbicka, A. (2009). "Reciprocity: An NSM approach to linguistic typology and social universals." *Studies in Language, 33*(1), 103–174.

Wilkinson, G. S. (1984). "Reciprocal food sharing in the vampire bat." *Nature, 308,* 181–184.

Wittgenstein, L. (1953). *Philosophical investigations.* Oxford: Basil Blackwell.

———. (2009 [1953]). *Philosophical investigations.* Translated by G. E. M. Anscombe, P. M. S. Hacker, & J. Schulte. Revised fourth edition by P. M. S. Hacker and J. Schulte. Oxford: Wiley-Blackwell.

Wivell, R. (1981). "Kairiru grammar." Unpublished MA thesis, University of Auckland.

Wordick, F. J. F. (1982). *The Yindjibarndi language.* Canberra: Pacific Linguistics.

Wrangham, R. W. (1975). "The behavioral ecology of chimpanzees at Gombe National Park, Tanzania." PhD dissertation, University of Cambridge, England, UK.

———. (2006). "Chimpanzees: The culture-zone concept becomes untidy." *Current Biology, 16,* 634–35.

Wyttenbach, R. A., May, M. L., & Hoy, R. R. (1996). "Categorical perception of sound frequency by crickets." *Science, 273,* 1542–1544.

Yamakoshi, G. (2001). "Ecology of tool use in wild chimpanzees: Toward reconstruction of early hominid evolution." In T. Matsuzawa (Ed.), *Primate origins of human cognition and behavior* (pp. 537–556). Tokyo: Springer.

Zeitoun, E. (2002). "Reciprocals in the Formosan languages: A preliminary study." Paper presented at the Ninth International Conference on Austronesian languages (9-ICAL), Canberra, 8–11 January 2002. Ms.

Zlatev, J., Brinck, I., & Andrén, M. (2008a). "Stages in the development of perceptual intersubjectivity." In F. Morganti, A. Carassa, & G. Riva (Eds.), *Enacting intersubjectivity: A cognitive and social perspective on the study of interactions* (pp. 117–132). Amsterdam: IOS Press.

Zlatev, J., Racine, T. P., Sinha, C., & Itkonen, E. (2008b). "Intersubjectivity: What makes us human?" In J. Zlatev, T. P. Racine, C. Sinha, & E. Itkonen (Eds.), *The shared mind: Perspectives on intersubjectivity* (pp. 1–14). Amsterdam: John Benjamins.

Index

Page references for figures are italicized.

About the Author

Anneliese Kuhle is postdoctoral researcher at the Institute for English Language and Literature, Freie Universität Berlin. She studied English linguistics and biology at the Freie Universität and, from 2009 to 2010, conducted a one-year exchange to the School of Culture, History & Language at the Australian National University, Canberra. In 2012, she defended her dissertation on the "semantic core" of reciprocal constructions. Subsequently, Kuhle's research has been focused on the relationship between language and thought and the relevance of material cultures of nonhuman primates with regard to the evolution of human language. Her research interests include linguistic relativity, language change, grammaticalization, distributed cognition, and prelinguistic behavior.

www.ingramcontent.com/pod-product-compliance
Ingram Content Group UK Ltd.
Pitfield, Milton Keynes, MK11 3LW, UK
UKHW020216290926
15470UKWH00001B/212

9 781498 561211